EARTH SHAPERS

EARTH SHAPERS

HOW WE MAPPED and MASTERED
the WORLD, from the PANAMA
CANAL to the BALTIC WAY

Maxim Samson

THE UNIVERSITY OF CHICAGO PRESS

The University of Chicago Press, Chicago 60637
© 2025 by Maxim Samson
Published 2025
Printed in the United States of America

34 33 32 31 30 29 28 27 26 25   1 2 3 4 5

ISBN-13: 978-0-226-84474-9 (cloth)
ISBN-13: 978-0-226-84475-6 (ebook)
DOI: https://doi.org/10.7208/chicago/9780226844756.001.0001

First published in Great Britain in 2025 by Profile Books Ltd.

Library of Congress Control Number: 2025932533

♾ This paper meets the requirements of ANSI/NISO Z39.48-1992
(Permanence of Paper).

Contents

Introduction

Do we choose the path that already exists, or create a new one instead? Sidestepping the obvious metaphor, my question refers more plainly to the innately human way in which we interact with our daily surroundings. We see traces of this engagement everywhere. Mere steps away from a pavement specifically provided for our daily perambulations runs a trail of bare earth, carved into the ground by countless feet, to undercut a corner that obligates a few extra strides. Walking through a park, another route might have been engraved into the grass to reach a sports field as directly as possible, allowing its athletic creators to save as much energy as possible for the pitch. And by a busy road is the evidence that hundreds of people have refused to take the risk of waiting for a traffic light to change, or for a driver to permit them to cross, for a little further along, a pair of bald patches engage in a staring competition across the greying tarmac.

Both in origin and in name, these 'desire paths' reflect how we interact with the world around us. Where we consider a feature of the existing landscape unfit for purpose, we either abandon it, or tweak it according to the Goldilocks principle, to ensure that it's just right. For most of us, the routes we forge are unintentional: seldom do we think hard about the new pathways we're incrementally creating when we trudge across the grass, even if some open-minded architects and urban planners now willingly inspect these routes in order to adapt their future projects to users' preferences.[1] Eventually a distinct route materialises, sealing the fate of

I

the land below, for the clearer the path becomes, the more people acknowledge its legitimacy as a route and vote quite literally with their feet on its future. Formerly a vestige of disobedience, a sign of defiance against inflexible design, the desire path now invites conformity. What had once been an organic expression of our human inclination to transform the planet according to our needs and desires now has the appearance of having always been there.

Desire paths, along with their deeply sunken counterparts, holloways, are micro-scale exemplifications of our broader desire to mould and remould our world so that it becomes more inter-connective – a phenomenon I call *earth shaping*. In keeping with geography's literal Greek meaning of 'earth writing', cultural geographers call attention to the notion that our planet's various 'cultural landscapes',[2] fashioned by humans onto the natural world, can be 'read' like story-filled texts.[3] Earth shaping adopts the same principles, but adds to them a specific emphasis on the manifold power of geographical connections. Human history has been written in geographical connection – and when you know what to look for, these stories, both obscure and renowned, are everywhere. This book explores the reasons why we engage with our surroundings through connection, and how, through our actions, we write ourselves and a very specific history into the ground.

This is an opportune moment to examine our planet's multi-farious linkages, for never has the importance of geographical interconnectivity been as palpable, nor so easily overlooked, as it is now. For millions of us, a day without accessing the internet is unthinkable, whether to read news stories from other coun-tries, connect with family, friends and colleagues without needing to leave our seat, or consult an email inbox filled with tiresome scams and spam from places we've never been. (By the same token, how many of us check that the accommodation we're planning for a trip offers Wi-Fi, particularly in rural areas and countries where internet access remains scarce?) Almost any item we see or touch in our homes has already taken a veritable odyssey to reach what still may not be the final destination in its roving life. And belatedly, considering that over two centuries ago the eminent Prussian geographer Alexander von Humboldt warned of the damage wrought by human practices on natural contexts and processes, more and more of us are cottoning on to the profound ways in which our actions affect the environment, whether locally or thousands of miles away. Interconnectivity is everywhere – and yet most of the time, we barely notice it at all.

Nor is our fashioning of geographical connections something new. Ancient trade routes such as those running from China to Greece (the Silk Road) and from southern Arabia to the Eastern Mediterranean (the Incense Route) remain legendary for their role in facilitating the exchange not only of the commodities that give them their names – and many others besides – but of culture, ideas and knowledge as well. From the trio of exotic gifts given to Jesus by the biblical Magi to the Buddha's wanderings across the Indian subcontinent, it is easy to take for granted the sheer extent to which our world and our narratives about its history have been shaped by connections. More often, our attention is drawn to issues of *dis*connection: border disputes, segregation, the hazards associated with crossing natural obstacles such as deserts, oceans and mountain ranges. Far less often do we think so deeply about our planet's many connective phenomena – except, of course, when they fail to work as expected.

In tandem with our widespread interest in the world's various boundaries, we are hard-wired to think in terms of discrete places, of countries and cities, hills and lakes, buildings, monuments, parks and more, which can inspire us, nourish us and allow us to situate ourselves on the planet. The linkages *between* them do not generally captivate us to the same extent, a truth that becomes apparent whenever we travel. Whether on our daily commute or to visit a distant land, we tend to be more concerned about specific endpoints than the fuzzy areas in between, which, notwithstanding any eye-catching areas of outstanding natural beauty, we barely register as anything more than a kind of ambiguous middle space. Yet the connection – the route we followed, in such instances – is part of what makes the places at either end meaningful, having enabled our journey and possibly even permitting the sites in question to have developed in the first place.

To return to desire paths, it would be too easy to classify them solely as markers of convenience. This certainly is a factor, but it's not the end of the story. By chiselling a physical pathway to expedite our journey between discrete places, we carve ourselves, our stories and our ideals into our surroundings. In reshaping and fusing the locations most relevant to us, we bend the planet to our will, writing our history on the landscape and leaving behind a continuously evolving palimpsest of memories, tales and philosophies. In time, the connections we etch can become so familiar, so normal, that we no longer even question their *un*natural origins. This is the power of earth shaping.

That we choose to mould a more meaningful and interconnected planet might seem prosaic or banal, but it does not mean it is unimportant, uninteresting or devoid of purpose – very much the opposite, in fact. It allows us to acquire resources, build alliances, increase our influence and rejuvenate damaged lands. From the forging of sustainable food supplies to the construction of effective irrigation channels, the reasons why civilisations have risen and fallen are profoundly tied to their ability to shape the earth in their favour for the long term. Few of us can live in complete isolation from the rest of human society, necessitating that we rework our environments to make them better connected, whether internally or with the outside world. At the smallest of scales, even desire paths bring intrigue: the social network Reddit features two separate bulletin boards devoted to showcasing the logical and, in most cases, hilariously lazy routes sculpted into the ground by shortcut seekers globally, while a third is dedicated to the signs and barriers erected by frustrated conformists in response. And, broadening our scope, desire paths have been fundamental to the modern history of the American West, among other places, these trails being blazed and followed by thousands of wagon-riding settlers in the nineteenth century, whose energy, ideas, faith and diseases have all, in one way or another, shaped the evolution of California, Oregon, Utah and more. Geographical connections are a key piece of our planetary puzzle: without them, the image remains fragmented.

In some cases, the connections we seek to produce are practical, pragmatic, logical. In others, connections are more conceptual, ideological or idealistic. Often, our connections are rooted in both types of interest, using the notion of networks to help fashion linkages that are more tangible. The potential advantages of earth shaping are as enormous as they are varied. There is one significant challenge, and one major problem, however: earth shaping is not monolithic, and it is rarely equitable. As a result, the connective vision sought by some is unlikely to be appreciated by all.

At first glance, Ras Lanuf is a thoroughly ordinary sort of place. Neat, beige homes and stocky palm trees line tidy streets, while a handful of banks and shops, mosques and schools serve the various daily needs of this small Libyan town's residents. A little further along the crystal-clear Gulf of Sidra, a modest commercial port conveys black gold far and wide, each ship provoking some periodic excitement in an area that, today, sees little action. Considering that Ras Lanuf's one obvious distinguishing characteristic – its enormous and lucrative oil refinery – made it a key battleground during the war-torn 2010s, the town's return to relative obscurity is a change that its residents most likely cherish. Beyond the refinery, the town certainly looks little different from many of the other settlements of Sirtica, a vast, barren region of ochre sands and scattered oil fields between the Mediterranean and the Sahara. Yet scratch beneath the surface and Ras Lanuf becomes improbably remarkable, a place whose past, present and future have been shaped and reshaped by various powers, seeking to forge competing conceptions of connectivity both imagined and real.

This area was first put on the map, literally as much as figuratively, in the fourth century BCE in a brief and tangential story of a diplomatic dispute told by the ancient Roman historian Sallust. Following a long period of warfare, the Carthaginians (from modern-day Tunisia) and the Cyreneans (from a Greek colony in what is now eastern Libya) sought to define a mutual border somewhere in the featureless desert plain between them. The unorthodox arrangement was that both parties would send a pair of runners from their respective capitals on the same day, at the same time. The point where they met would mark the location of the new border. Unfortunately, this agreement quickly proved controversial: the Carthaginian team of twins managed to proceed considerably further than their Cyrenean counterparts, who accused the former of cheating. Unwilling to yield, the Carthaginian brothers ultimately agreed to be buried alive at the meeting point. To honour such a show of patriotism and bravery, which allowed Carthage to claim a much larger territory than the losing Cyreneans had anticipated, a pair of altars was constructed on the spot where the twins were buried, on the outskirts of what is now Ras Lanuf.[4]

Whether Sallust's tale was true, apocryphal or somewhere in between, it proved irresistible over two millennia later to the region's dominant power in the 1930s. Under the fascist leader Benito Mussolini, Italy was a nation deeply intrigued by the opportunities geographical connection can provide, first occupying and unifying the historic provinces of Tripolitania (a region that had once been a significant part of the Carthaginian Empire) and Cyrenaica as a single entity,* and then tacking this newly

---

* Although both of these ancient provinces had already been claimed by an array of foreign powers over the centuries, until the Italian occupation they were nevertheless treated separately from one another administratively and politically.

integrated 'Libya' conceptually onto its self-proclaimed empire. With the territories attached and traditional borders overcome, a triumphal arch was promptly erected in the very location where the Carthaginian altars, by now long since disappeared, had once stood.[5]

Superficially, one might think that the arch emphatically honoured the two brothers, being named for these Philaeni* and depicting the twins both as bronze statues and on bas-reliefs.[6] However, other aspects of the arch's design were chosen to ensure that Rome would be at the forefront of viewers' minds, a hazardous Mediterranean voyage away though it was. The architect Florestano Di Fausto constructed the arch from the same white travertine stone as that found throughout the Eternal City, thereby drawing a conceptual link between North Africa and the ancient Roman capital. The arch's decorative elements went a step further, with a carved panel depicting Mussolini saluting the King of Italy, Victor Emmanuel III, in front of Rome's seven hills, and an unmissable inscription attributed to the Roman poet Horace, declaring (in Latin, of course), 'O fostering Sun, may you never see anything greater than the city of Rome'.[7] Though some observers drew comparisons of their own – British troops fighting in the Second World War's North African campaign likened it to London's Marble Arch[8] – for Italy, this monument was both a physical manifestation of its imperial might and a symbol of its ability to join new territories to the motherland.

In writing a distinctly fascist worldview onto a remote part of the North African landscape, the arch was therefore critical to returning Rome – and by extension, Italy – to some of its ancient dominance in the Mediterranean. Whereas the ancient Carthaginians had sought to divide themselves from enemy Cyreneans, the modern Blackshirts embraced connection, reappropriating the

---

* This Greek term, meaning 'lovers of praise', suggests that at least one of Sallust's sources was more than a bit disgruntled at their achievement.

Philaeni story in the process of reviving their forefathers' practice of building arches to celebrate the incorporation of new territories, including in North Africa. Still, Italy's connective vision was not merely symbolic or conceptual. To reinforce the sentiment that, in one way or another, all roads lead to Rome, the colonists additionally conceived another engineering project to impose upon the landscape.

Eventually stretching nearly 2,000 km (around 1,200 miles) from Tunisia in the west to Egypt in the east, the Libyan Coastal Highway would allow Italy to replace slow, infrequent steamship voyages with a modern, militarised transport artery,[9] and provide a logical axis for all sorts of auxiliary connective infrastructure in the region, including aqueducts, embankments and pipelines. Through setting thousands of its liraless natives to work in North Africa, Italy regarded this earth-shaping project as more than just a practical tool of 'civilisation', by which barren lands would be converted into useful, cultivable, integrated territory.[10] The road's status as an infrastructure project of some repute also helped Il Duce 'justify' the new Libya's incorporation as a fully fledged Italian province, a 'Fourth Shore' connected emotionally and officially with the lands along the Tyrrhenian, Adriatic and Ionian coasts to the north.[11] Reinforced by the efforts of Italian archaeologists in excavating ancient Roman sites such as Leptis Magna along the highway, and sending their treasures to museums back home, the feeling, on the European side of the Mediterranean at least, was that Libya's bonds with Italy were historic and profound.

Such connections do not tend to be fully consensual, however. Although many locals quickly came to view the arch specifically as a symbol of Italian imperial power – as Mussolini had intended – they detested it for the same reason that the fascist leaders prized it. It is certainly no accident that shortly after his successful coup d'état and revolution in 1969, Libya's long-time authoritarian leader Muammar Gaddafi dynamited this monument to smithereens. A visitor to Ras Lanuf can no longer see the landmark that briefly distinguished this otherwise little-known location (though the bronze statues and fragments of the bas-reliefs did manage to survive and have since been preserved at a nearby museum).[12]

The highway, on the other hand, has not merely survived; it has become fundamental to the country's continual vicissitudes and evolution, integrating it as one while inviting discord and tension ever since. Independence, achieved little more than a decade following the road's completion, has seen different leaders retain the name 'Libya' and the borders created by Mussolini. Whereas most of the Italian-style villages the highway once connected have gradually been abandoned,[13] the road itself has been widened and repaved, allowing the old colonial objective – replacing Indigenous nomadism and subsistence pastoralism with fixed settlement and private property ownership – to be realised.[14] Even in the new millennium, the legacy of the road remained so influential that Gaddafi convinced Italy to build his country a *new* highway untainted by colonial oppression, in return for his support in curtailing migrant flows from Africa to Europe. However, before construction could begin on this modern transport link, the outbreak of civil war saw the original route's own purpose morph instead. It became a key conduit for people seeking to either join or flee violence, a popular trafficking route and (particularly close to refineries and ports) a key arena of conflict itself.[15] By April 2019, amid a military offensive launched by the eastern military commander Khalifa Haftar to capture Libya's western region, the writing was on the wall: the highway once conceived as Libya's physical nexus and conceptual axis was now so embroiled in strife, so broken, that it needed to be closed. More than a year would pass before Libya's 'essential lifeline' connecting west with east opened again.[16]

The arch is no more, and the highway, damaged, repaired and enlarged since the 1930s, looks very different now from in

9

Mussolini's time. Today, the scars of Libya's recent conflicts represent the most conspicuous layer in the region's intricate palimpsest, inscribed two and a half millennia ago by the Carthaginians and progressively complicated by North Africa's metamorphosing powers and their respective geopolitical concerns. The connective relevance of the road once created to unify Italy's North African territories has evolved: it has been reworked among other European-created road clusters into a post-colonial, pan-African highway tying west (Senegal) with east (Egypt), emotionally as well as practically.[17] Although it is not currently possible to travel as far as Dakar, owing to the long-term closure of the Morocco–Algeria border, from Ras Lanuf one can feasibly drive far further than the Italian colonists had anticipated. Like many other places across the world, this small coastal town may appear entirely mundane at first glance. That is, until we consider its relationship to other links in the chain. Every landscape tells a story – the challenge is knowing where to look.

Each of the chapters in this book presents a distinct means by which we use the power of geographical connection to shape and reshape the world. They span the globe both geographically and temporally, yet each shares a theme of humans seeking to tweak their surroundings in order to write their worldview onto the earth. However, earth shaping is never apolitical, and its impacts are often far-reaching. Just as these connections have all profoundly shaped local landscapes, creating ramifications and legacies that transcend borders, they have simultaneously been confronted by opposition, whose visions of connectivity can differ quite significantly.

We will explore geographical interventions designed to build *order*, a truth known by the ancient Romans in their expansive road network. Less well known but no less impressive was the remarkable road system developed by the Inca in the Andes, which cut across some of the planet's most forbidding terrain. In

Chapter 1 we will meet this fastidiously maintained and distinctly hierarchical network, the Qhapaq Ñan, which was fundamental to enlarging and managing the Inca's domain. The road system additionally wrote a series of conceptual connections onto the landscape, structuring both Cusco and the empire itself, Tawantinsuyu, into four quarters according to invisible ritual pathways. Ultimately, though, Tawantinsuyu was crippled by its road system's extent, efficiency and engineering excellence: Spanish conquistadors harnessed this extraordinary feat of geographical intervention for their own ends, capitalising on its functional advantages to quickly reach Cusco, plunder its resources and decimate this short-lived empire.

In other cases, *extraction* can be built into the landscape from the ground up, as was the case with Mozambique's railways, which we will explore in Chapter 2. Portuguese colonisers conceived these routes as a means of conveying Southern Africa's mineral wealth from the interior to the coast, while sending cheap labour the other way. However, the Portuguese weren't alone in recognising the economic and geopolitical potential of railways, and in a short time, the routes intended to legitimise their sovereignty were fashioned according to their colonial rivals' interests instead. From a brutal civil war after independence, to ongoing disputes concerning the power of foreign industries, Mozambique's past and present have been profoundly shaped by the distinctly fragmented assortment of railway networks left behind by European colonisers, whose primary purpose remains extraction.

One of the more familiar ways we interact with the world around us is in the name of *convenience*, and nowhere better epitomises this than the Panama Canal, the subject of Chapter 3. Arguably humanity's most impressive infrastructural achievement, a water connection between two oceans whose development impelled a nation's (quasi-)independence and whose administration strained relations between its communities, this canal has become essential to modern life across the planet. Indeed, the

Panama Canal, an aquatic desire path of sorts, more than simply accelerated the global transportation of goods. Its completion under US command additionally marked a paradigm shift in the global power balance, in which a new, technologically advanced superpower would position itself at the focus of the world's attention.

The United States' claim to being the planet's figurative and geographical centre has remained robust more or less ever since. But one of the contemporary world's most ambitious nations is striving to become a realistic rival. Chapter 4 explores how Saudi Arabia is using the power of geography to *reimagine* our planet and our relationship with it, concentrating on its flagship 'smart city' THE LINE. Not only does this city seek to redefine Saudi Arabia as a hub of talent and innovation at the crossroads of Asia, Africa and Europe, but in championing a profoundly interconnected design, THE LINE also demands that we rethink what a city even is and looks like. Both lustrous idyll and Orwellian dystopia, this controversial city represents an intriguing case of how imagination and ambition, along with physical and conceptual connection, can fuel a rebranded and refashioned landscape.

A different way in which geographical connections can spark the imagination rather than being solely tangible provides the theme of Chapter 5: *resistance*. We tend to think of geography as distinctly physical elements of the world around us, and in the case of resistance, this is as true of military tunnels in Vietnam or Gaza as the interconnected basements used by Prohibition-dodging bootleggers in Moose Jaw. However, in other cases, connective resistance networks assume a more conceptual form, such as the Underground Railroad tying together the safe houses used by enslaved African Americans to flee northwards in the nineteenth century. This chapter's case study combines elements of the two: it was conceived in secret and its physicality was unusually ephemeral. Yet the Baltic Way's (or Baltic Chain's) effectiveness owed much to its conspicuously public manifestation, a human chain tying together three nations whose principal commonality was their opposition to the Soviet regime. On top of enabling Estonia, Latvia and Lithuania to wrest themselves from foreign occupation, the Baltic Way enabled these historically dissimilar states to view their pasts and futures as intertwined and, correspondingly, to be regarded as a continuous geographical buffer against Russian aggression. This protest – thanks to which new alliances were created, freedoms were achieved and frontiers were drawn – encapsulated the power of geographical connectivity in remoulding a planet whose future is malleable rather than fixed.

A concern with the future is additionally relevant to the many geographical interventions that seek to *restore* damaged environments and regions. The case study presented in Chapter 6, the Great Green Wall, epitomises geography's potential to inspire collaboration in the face of multiple and interrelated environmental, social and political challenges. Having originally been conceptualised as an uninterrupted tree buffer stretching from the Atlantic Ocean to the Red Sea, this initiative now necessarily defines itself more loosely as an interconnected mosaic of re-greened and productive territory at the Sahara's southern edge. However, despite some important successes in rehabilitating arid lands, a fateful melange of desertification, poverty, terrorism and conflict continues to forestall the development of the very project conceived as a panacea to these problems, and magnifies the challenges of tackling issues that, like the initiative itself, transcend national borders.

The Great Green Wall elucidates the potential of geographical connections in protecting our fragile planet, all while the same region's long-established trading routes are exploited by smugglers and extremists for more malignant purposes. It reflects how earth shaping does not assume a single form, and how just as geographical connections can be moulded carefully to preserve the environment, they can alternatively be modified much more substantially for quite different ends. Chicago, the focus of Chapter

7, presents an informative case of the same geographical connections cherished by some, being *co-opted* and manipulated beyond recognition. Here, settlers profited from Indigenous knowledge of the American Midwest's ridges and waterways to first seize and then reshape a gently managed natural environment into a bustling industrial metropolis. To this day, the battle for recognition persists in a city whose Indigenous origins and ways of life – which were informed by a spiritual concern with the interdependence between humans and nature – were quickly obscured by one of the most conspicuously systematic forms of earth shaping in history.

The substitution of Indigenous understandings of earth shaping for a distinctly modern approach rooted in materialism and control might suggest that traditional beliefs are destined for marginalisation in the contemporary world. However, across the planet, millions of people continue to regard powerful metaphysical connections between specific places as sources of *vitality*, from the straight, invisible ley lines allegedly running between a miscellany of prehistoric and historic landmarks,* to India's Char Dham, a Hindu pilgrimage circuit joining four sacred sites aligned in a neat diamond shape. In Chapter 8 we will consider the Baekdu-daegan of Korea, which embodies the enduring influence of traditional beliefs in reshaping our world. This chain of mountains directing the distribution and configuration of the peninsula's settlements provides an unlikely source of shared national pride for the planet's least friendly neighbours. Reinforced by leisure activities and political actions that emphasise its cross-border bonding

---

* Although the originator of this theory, the English amateur archaeologist Alfred Watkins, had suggested that such lines simply represented ancient trade and navigational routes, the notion has since become better associated with esoteric movements that claim they demarcate mystical earth energies and can be used to guide and communicate with extraterrestrials – a form of connection many others deem far-fetched.

potential, the Baekdu-daegan provides much-needed hope that someday, Korea can become one again.

Encompassing different continents and addressing a range of pertinent issues, through these eight chapters we'll see time and again how we humans mould our surroundings, feeding a perpetual desire and determination to manage our complex world as we see fit. For geography is never an inevitability, and connection is rarely innocuous. The story of Ras Lanuf typifies how a humble place can become significant due to the multiple, consequential types of connectivity, ideological as well as pragmatic, in which it is bound. The essence and evolution of the locations I've chosen to focus on are similarly tied to a framework of connection. Political and potent, grounded and germane, these connections all illustrate the uniquely human ways in which we engage with our planet and our dedication to adjusting it according to our perceptions and priorities. It's never sufficient to think about a place in a vacuum, and it's wrong to assume that our relationship with the world is inexorable. Looks can be deceiving, obligating us to delve ever more deeply into our surroundings and the processes and choices behind their appearance: for by examining geographical connections, we become better able to control the present and direct the future.

1

Order: The Qhapaq Ñan

---

*In the memory of people I doubt there is record of
another highway comparable to this, running through
deep valleys and over high mountains, through piles of
snow, quagmires, living rock, along turbulent rivers;
in some places it ran smooth and paved, carefully
laid out; in others over sierras, cut through the rock,
with walls skirting the rivers, and steps and rests
through the snow; everywhere it was clean-swept
and kept free of rubbish, with lodgings, storehouses,
temples to the sun, and posts along the way.*

Pedro de Cieza de León[1]

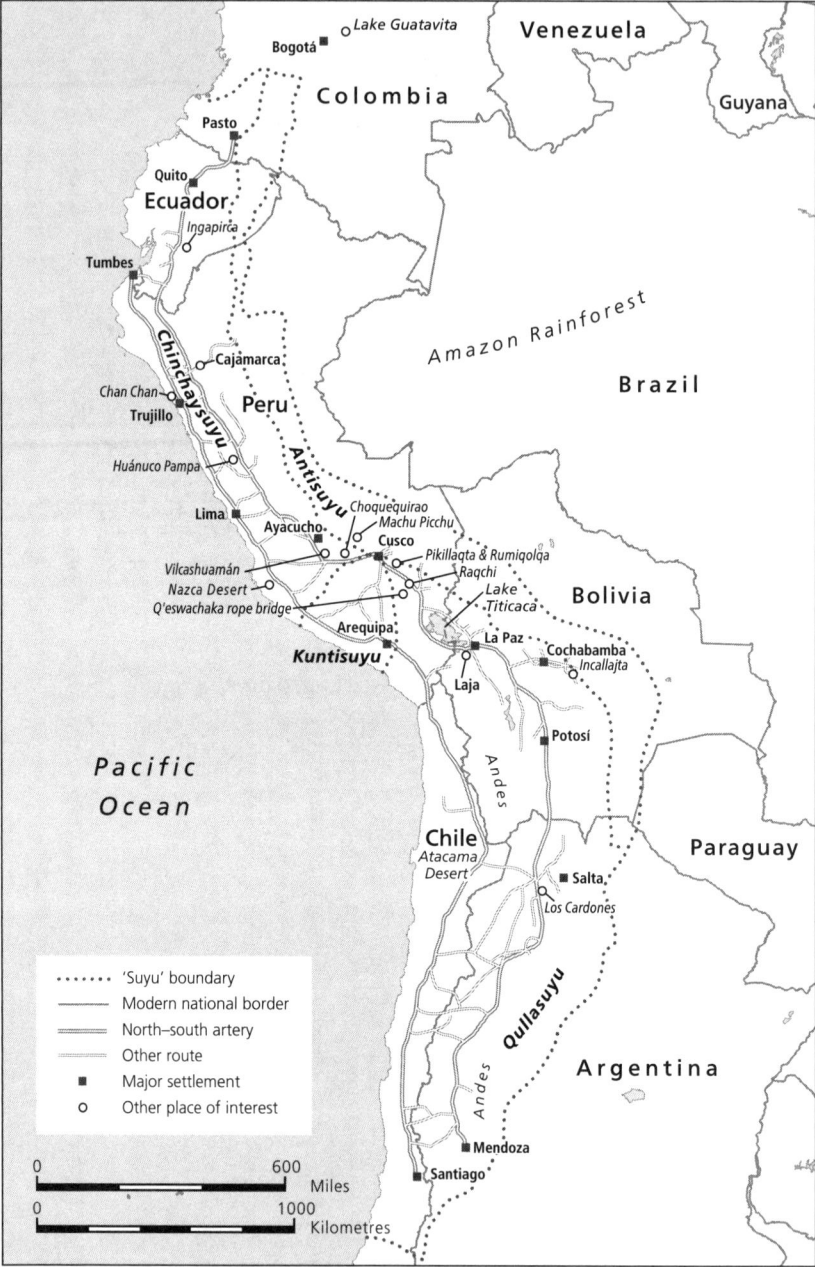

Venezuela

Guyana

Colombia

Lake Guatavita

Bogotá

Pasto

Quito
Ecuador
Ingapirca

Tumbes

Chinchaysuyu

Cajamarca

Peru

Chan Chan
Trujillo

Huánuco Pampa

Lima

Antisuyu

Choquequirao
Machu Picchu
Cusco
Ayacucho *Pikillaqta & Rumiqolqa*
Vilcashuamán *Raqchi*
Nazca Desert *Lake Titicaca*
Q'eswachaka rope bridge

Amazon Rainforest

Brazil

Bolivia

Arequipa
Kuntisuyu
La Paz
Laja
Cochabamba
Incallajta

Potosí

Pacific Ocean

Andes

Chile
Atacama Desert

Paraguay

Salta
Los Cardones

Qullasuyu
Andes

Argentina

Mendoza
Santiago

・・・・・	'Suyu' boundary
——	Modern national border
≡≡≡	North–south artery
⋯⋯	Other route
■	Major settlement
○	Other place of interest

0   600
Miles

0   1000
Kilometres

There are very few places in the world that inspire as much awe as the western portion of South America. In Colombia, famed for its verdant and vertiginous coffee plantations, the Sanctuary of Las Lajas may be the continent's most striking pilgrimage destination, its Gothic Revival arches and statues of angels extending across a deep river canyon (rivalled only, perhaps, by the Zipaquirá Cathedral, built deep in a salt mine and illuminated with purple lights). In Ecuador, combining a perfectly conical shape with a snow-capped peak, steep Cotopaxi is for all intents and purposes the quintessential volcano, while a short drive away, reclusive spectacled bears and cacophonous cocks-of-the-rock are just two of the many species that render the country's cloud forests among the planet's most biodiverse environments. In Peru, the Nazca Desert's ancient and enigmatic geoglyphs continue to elicit theories both plausible and fanciful as to their purpose and origin, while the hues of the kaleidoscopic Vinicunca are more majestic even than its alternative names, Rainbow Mountain or Mountain of Seven Colours, would suggest. Spreading over the southern border into Chile, the bone-dry Atacama Desert, whose orange and red soils are warmer than the air above, presents other-worldly landscapes befitting a place long used by NASA to test its Mars rovers. Meanwhile to Peru's south-east, Bolivia's Uyuni salt flats, etched with natural polygons and, on occasion, gleaming like an endless horizontal mirror, are as breathtaking as their altitude of over 3,600 metres (about 12,000 feet) above sea level. And further south, *gastrónomos* may be interested in knowing that Argentina's Mendoza region is now one of the world's foremost wine producers, while Salta boasts more than just some of the country's finest architecture from the Spanish colonial period; it is also famous for its succulent and spicy empanadas.

Yet though these illustrious archaeological sites, spectacular vistas and grand cities are all highlights of South America, beneath – or rather, between – them all is a feat of earth shaping so exemplary that it astounded the Spanish conquistadors and allowed a small Indigenous kingdom to become perhaps the best-administered empire of all time. Romanticised by some as the civilisation that built Machu Picchu and domesticated cheery llamas and alpacas, and deprecated by others as a brutal people with a penchant for sacrificing children and recycling enemies' body parts into cups and drums,[2] the Inca were exceptional in their recognition that geographical connection can bring order to the human and natural worlds alike. In under a hundred years, a non-literate civilisation from Cusco that didn't use wheels, arches, maps, machines or currency and couldn't smelt iron demonstrated that roads can be used as a tool for formalising control over a multitude of previously fragmented regions and rivals across a uniquely extreme assortment of natural landscapes.

Stretching for 40,000 km/25,000 miles (and almost certainly far more) along the world's longest above-water mountain range and the highest outside Asia, this road system, the Qhapaq Ñan, connected the planet's largest rainforest, the Amazon, with its driest hot desert, the Atacama, and its tallest volcanoes, such as Ojos del Salado, with its biggest ocean, the Pacific. In fusing the four *suyus* (regions) of their empire Tawantinsuyu, the Inca's 'great road' or 'way of the powerful' materialised their distinctive integration of engineering and spirituality, and demonstrated their capacity to oversee every aspect of their domain.[3] That is, until that same network's very calibre was used against them, bringing the short-lived empire swiftly to its knees.

Half a millennium on, the Inca are long gone and most of the Qhapaq Ñan has been lost to time. Yet from ethnicity to socioeconomic status and from national symbols to sacred spaces, the identities of all six of the countries to inherit the lands it joined and the lives of millions of their inhabitants are still linked both directly and indirectly to pre-Columbian America's greatest infrastructure accomplishment. South America's modern borders conceal the meaningful interrelationships between sites from Colombia in the north to Chile and Argentina in the south, but written into the landscape, countless stories of ritual and trauma, settlement and abandonment survive, all facilitated or provoked by a power-laden, pedestrianised road system. As part of the basis of Andean geography, history and spirituality, the Qhapaq Ñan beseeches us to look more deeply, to acknowledge that the connections between iconic places can matter at least as much as the places themselves. Tawantinsuyu's notoriety may be incommensurate with its longevity, but its comprehensive legacy lives on in its road network, which is instantaneously perceptible, should one scratch beneath the surface.

The Qhapaq Ñan was an Inca achievement, but just as with any other imperial power throughout history, this civilisation did not start its work on a blank canvas. Though the Andes present numerous obstacles to infrastructure development, various prior communities had already proved that through ingenious engineering techniques appropriate to local landscapes, it was possible to develop thriving civilisations.

For 600 years from approximately 100 to 700 CE, the Moche along northern Peru's freshwater-deprived coastal desert had dug deep trenches as irrigation canals to divert river water to fields. Their successors in the region, the Chimú, then replaced these channels with contour canals that maintained a slight but constant slope to redirect water as far as 70 km (over 40 miles) away.[4] In fact, for the Chimú, these water connections were just one part of a sophisticated irrigation system, comprising sunken gardens called *huachaques* (which allowed farmers to work the moist soil beneath the dry topsoil), walk-in wells to tap groundwater and major reservoirs for storage.[5] As a consequence of their outstanding surveying and engineering skills, before their eventual conquest by the Inca in around 1470, this moon-worshipping civilisation had also managed to construct the largest city in the Americas and the biggest adobe settlement on the planet, Chan Chan.[6]

Another Indigenous civilisation, the Wari, left its mark on Peru's southern highlands several centuries before the Inca emerged, constructing the walled settlement of Pikillaqta close to the future Inca capital, Cusco, and – even more helpfully to the latter – what was likely an aqueduct at Rumiqolqa. According to local legend, this was developed as part of a competition between two suitors to the beautiful daughter of a Wari leader, who, anxious about her people's future subsistence, demanded that the young men first prove their ability to provide the community with a reliable water supply.[7] The Inca later appeared to adopt and expand this impressive roadside structure, likely using it as part of a regional canal system as well as both quarry and gateway to their capital.[8]

Further south-east over the contemporary Bolivian border, for centuries the Tiwanaku established their own civilisation near Lake Titicaca, reworking hillsides into step-like *andenes* terraces to minimise soil erosion, reduce frost damage and control the amount of water available to crops such as maize, and constructing *waru waru* cultivation and irrigation systems, characterised by alternating embankments and canals.[9] Though the reasons for their downfall, also around 1000 CE, remain disputed, they, like the Moche, Chimú, Wari and other groups in modern-day Peru and Bolivia, laid the literal, crucial groundwork for the Inca, creating trails and roads convenient for expansionist purposes.

Indeed, just as the Wari had constructed roads to stretch their empire from their eponymous capital near Ayacucho in south-central Peru both south-east towards Lake Titicaca and far further north-west to Chachapoyas and Piura,[10] the Sapa Inca (Emperor) Pachacuti recognised that this infrastructure represented the most effective means of first claiming and second administering distant lands. Having ascended to the Inca throne in 1438 unexpectedly, fending off an invasion by the Chanca after his terrified monarch father and presumptive heir older brothers fled, Pachacuti quickly initiated a programme of growing Cusco into a road-based empire.[11] Comprising two main north–south highways inter-connected by various shorter, transversal west–east roads, this ladder-shaped network was like nothing South America had ever seen. The western route largely followed Peru's coastal plains, before deviating further inland through the Atacama Desert to hug the foothills of the Andes at least as far as Chile's capital, Santiago. Meanwhile, the eastern route ran through the undulat-ing grasslands of the Central Andean plateau before dividing to circle Lake Titicaca.[12] Other than this relatively short split, this highway was otherwise largely continuous from Pasto in southern Colombia all the way to Mendoza in western Argentina – a dis-tance of approximately 6,000 km (3,750 miles), not dissimilar to that between Lisbon and the Caspian Sea, but at an average alti-tude of 3,000 metres (10,000 feet) above sea level (and sometimes climbing as high as 5,000 metres/16,000 feet).[13]

Whether by enhancing existing routes or building new ones, the Inca always ensured that their roads were adapted to diverse natural environments instead of following a one-size-fits-all tem-plate. In desert regions, short walls prevented sand from caking roads that were built low to be close to essential water sources, while in wetland regions such as around Lake Titicaca, earthen causeways and embankments provided protection from fluctuating water levels. In highland areas at risk of landslides and flooding, roads were typically constructed along upper valley slopes and featured culverts and drains to reduce water run-off, retaining walls to increase stability, and paving stones or cobbles to provide a more uniform surface. Though the Inca typically preferred to construct straight roads – as with the Recta del Tin Tin, which crosses the high-altitude desert of what is now Argentina's Los Cardones National Park – they, unlike San Francisco's planners centuries later, recognised that steeper slopes necessitated some adaptability on the part of humans, resulting in narrower, winding roads and even flights of stairs. In any case, wherever possible, the Inca tried to bypass those environments most dangerous to travellers: their routes skirted the edge of the Atacama along the Andean foothills, where far more water is available, and ascended far above major rivers in highland areas.[14]

In addition to being excellent road builders, the Inca were outstanding bridge engineers, enhancing their growing empire's connectivity. Recognising that their rope suspension bridges across narrow gorges represented critical choke points during times of war, they deployed full-time overseers to guard the longest crossings, many of whom would weave new cords from wild grasses while on duty to pre-empt potential sabotage.[15] Although the Inca also built a relatively small number of stone-lined culvert bridges over short, shallow stream crossings,[16] the fragility of other types of crossings happened to offer certain advantages to the Inca regime. For instance, the regular maintenance and replacement of a rope bridge acted as a social glue for the communities on either side and theoretically reduced the risk of rebellion,* while log bridges, a good example of which stood on the edge of Machu Picchu, could easily be removed should a threat be detected.[17] Even more hair-raising than these crossings – especially rope

---

* There is no better evidence of these crossings' power to create social and symbolic as well as physical connection than the fact that the residents of four different Indigenous communities still renew the Q'eswachaka rope bridge south of Cusco for free. Spanning a 30-metre (100-foot) gap across the Apurímac River, this impressive construction collapsed in March 2021 due to a decline in maintenance during the COVID-19 pandemic, prompting teams of workers on either side of the ravine to immediately re-weave it.

bridges, which sagged in the middle and would sway with the slightest wind, spurring terrified Spanish conquistadors to crawl on their bellies[18] – were *oroyas*, hanging baskets used to transport two or three people together over larger ravines.[19] Regardless of type, these bridges were essential to the Inca mission to reach and link otherwise detached areas, and in a reflection of this society's meticulous level of organisation, an official inspector (the *chaka suyuyuq*) was appointed to ensure that every link in the imperial chain was consistent with state policy.[20]

Certainly, to the Inca, the Qhapaq Ñan served as an instrument of control, bringing order to the natural world and miscellaneous populations alike. New territories were either annexed through diplomacy and elite marriage alliances or conquered by force, and were quickly connected to the existing road network. This military-political strategy symbolically indicated Inca domin-ion, while facilitating the movement of troops to quell rebellions, prepare for further advances and support civil works. Vanquished rivals were intentionally dispersed to distant parts of the empire, and their roads promptly connected with lands these newly dis-placed people had never seen and from which they would never be allowed to return. Formerly distinctive and multitudinous Indigenous communities were shuffled throughout the growing empire to engineer a form of social integration that served to mitigate opposition along ethnic lines – a strategy whose legacy today includes the significant cultural diversity of cities such as Cochabamba in Bolivia, whose valley the Inca used for the mass cultivation of maize and coca. Almost invariably practical and resourceful, the Inca tended to forgo permanent garrisons (as well as city walls), instead periodically moving their troops along the Qhapaq Ñan whenever a threat was identified or suspected.[21]

Building order into Tawantinsuyu involved far more than constructing a militarised road network and using it to alter its provinces' demographic make-up, however. The Inca elite addi-tionally instituted a highly regulated political, social and economic system, of which the Qhapaq Ñan was a key component. Though they exerted control differently throughout the empire, consistent in Inca policy was a phenomenal level of standardisation, whereby each person's role was carefully defined and affected how they used the roads – if at all.[22] Of course, the Qhapaq Ñan was used by the Sapa Inca; the relevant roads were brushed clean before he and his gigantic entourage passed through.[23] But much further down the social hierarchy, and hardly less important from a practical perspective, couriers called *chaskis* rushed between official relay stations with small packages and *khipus*, coded, knotted strings that enabled these enviably fit runners simultaneously to play a giant game of telephone and record their accounting without needing to write anything down. Although seemingly primitive in comparison to modern mail transport – lost postcards and birthday presents notwithstanding – this system was suited to traversing some of the planet's most challenging terrain and was impressively fast and efficient, relay teams being capable of covering 240 km (150 miles) in a single day. Still, beyond the most privileged classes and *chaskis* as well as the army, llama caravans, porters and authorised pilgrims, traffic on the Qhapaq Ñan was carefully managed, being prohibited for subjects unless they had been sent by political elites for a special purpose, intensifying Tawantinsuyu's social distinctions as a result.[24]

Two groups given occasional dispensation to take the Qhapaq Ñan were *mitimaes*, colonists involuntarily transplanted to introduce Inca customs to new territories and thereby erode alternative loyalties, and labourers who were part of the *mit'a*. The latter was a turn-based mandatory service system which saw workers drafted to fulfil all sorts of duties required by the state, including farming, mining, constructing temples and fortifications and, of course, building and caring for the empire's roads and bridges. In a society that eschewed private commercial exchange and taxes, and prohibited the common classes from owning property, these workers acted as bona fide jacks-of-all-trades, taking it in turns to produce goods and provide public services throughout the empire, and in return receiving a stipend of textiles, food and *chicha* (sacred corn beer), as well as invitations to periodic feasts. Thanks to this pragmatic system, which transformed a longer Andean tradition of reciprocity called *ayni* into a patron–client relationship, the Inca elite was able to reinforce its own power: conscientiously maintained roads guaranteed the state-controlled circulation of workers and resources, while the empire's diverse and scattered populations were assured a consistent, sustainable supply of daily necessities as well as military leadership and political stability.[25]

Always conscious of the need for order and efficiency, the Inca leadership additionally oversaw the construction of essential services along the Qhapaq Ñan, from new administrative centres, which conducted complete censuses of new territories so that plans could be made for their infrastructure and allocation of *mitimaes*,[26] to fortresses (*pukaras*), which acted as military bases while distinctly conveying to local inhabitants who was in charge. At carefully measured intervals along the roads were *tambos*, inns offering travellers much-needed rest and refreshment.[27] Accompanying these structures or established separately near larger settlements and state farms were *qollqas*, warehouses storing grains, maize, clothes and weapons for soldiers in the event of war, as well as an assortment of fine crafts and ritual objects, all recorded carefully using *khipus* so that they could be replenished or distributed as and when necessary.[28] Although Inca authority wasn't absolute – highwaymen called *pomaranra* disrupted traffic, and it is plausible that in more remote locations, people would have accessed and used the roads in ways that didn't conform to official expectations[29] – practically, administratively, militarily and diplomatically the Qhapaq Ñan was fundamental to the empire.

In short, the Qhapaq Ñan was crucial to reproducing both hard and soft power. It enabled the Sapa Inca to receive information on every part of Tawantinsuyu and swiftly mobilise and move his troops whenever a disturbance was reported.[30] It allowed him to keep tabs on the circulation of every commodity, both functional and ritualistic, within the empire's borders.[31] It ensured that Inca rules and customs reached even the most far-flung places and defiant peoples, and through resettlement, it turned potential rebels into productive workers and cultural role models.[32] Living in Tawantinsuyu was presented as a privilege, its frontiers being described as the boundaries between the civilised and savage worlds, and the quality of its road infrastructure accordingly used to manifest Inca superiority over vanquished rivals.[33] In a part of the world characterised by natural geographical obstacles to population movements – lofty mountains, geologic hazards, tropical forests, arid deserts, scarce oxygen – everywhere and everyone was connected, operating as part of an integrated system that superseded ethnic difference and enabled travel which, though strenuous, was possible in a way it hadn't been previously. The Qhapaq Ñan's significance to every element of Inca society was unmistakable: around Lake Titicaca, previously dispersed communities now tended to live within half a kilometre (a third of a mile) of the system,[34] while in border regions, clean, orderly roads both facilitated and legitimised Inca control over a growing array of people and places. And the whole time, the roads supplied Cusco with the resources and expertise it needed to expand as the nexus between the empire's four *suyus*: Chinchaysuyu in the north, Qullasuyu in the south, Kuntisuyu in the west and Antisuyu in the east.*

Part of what made Cusco so irresistible as a nexus was the fact that its importance as a physical hub was intertwined with its

---

* Today, these provinces roughly correspond, respectively, to most of Peru and Ecuador and the south-western corner of Colombia; south-eastern Peru, south-western Bolivia, northern Chile and north-western Argentina; south-western Peru; and eastern Peru and north-western Bolivia.

role as an intersection of a more conceptual sort, allowing it to bridge the material and supernatural worlds. It is crucial to recognise, after all, that the long-standing Eurocentric propensity to distinguish sacred activities and spaces from 'ordinary', everyday life makes little sense to Indigenous Andean people. To them, the Qhapaq Ñan was conceived as simultaneously functional and ritualistic, integrating and connecting everything and everyone to achieve completeness and order.[35] To this end, the Inca shrewdly reappropriated and tweaked an older pan-Andean creation story so that it would explain their own emergence as a civilisation and ingrain a connection between the mythical and the physical.

Whereas in the original creation myth, the creator god Viracocha rose from Lake Titicaca to fashion the sun, moon and stars from its islands, in one of the most popular Inca origin myths, Viracocha's son, the sun god Inti, commanded two of his children to civilise and bring order first to this area, and later, by taking the main road to the north-west, to search for a fertile valley where his golden staff would sink readily into the earth.[36] Inevitably, this site was Cusco, with the effect that their existing capital and Lake Titicaca – a site of wider Andean significance – were intimately linked to each other both historically and in the cosmological geography the Inca sought to propound. Regarding themselves as direct descendants of the world's origins in this way, the Inca believed that they enjoyed divine permission to establish their own empire in the Andes,[37] and (unintentionally) set in motion a far longer legacy of depicting the sun on South American national emblems, including the current flags and coats of arms of Argentina, Bolivia, Ecuador and Uruguay, and historically Peru as well.[38]

Another key aspect of Inca belief was a three-part division of the cosmos into a celestial world (*hanan pacha*), a terrestrial world (*kay pacha*) and an underworld (*ukhu pacha*).[39] Though distinct, these worlds were crucially connected by conduits, including lightning, caves and *ushnus* – stone, stepped, pyramid-shaped structures, where carefully scheduled and choreographed rituals allowed the earthly realm to be joined with celestial bodies above and water below.[40] Whereas some caves and labyrinthine *chinkana** tunnels beneath Cusco may pre-date the Inca, their construction of *ushnus* encapsulated their additional, horizontal understanding of connection and its political as well as spiritual function. These ritual sites acted as patent symbols of Inca control over new territories, each connected by the Qhapaq Ñan to the most important *ushnu* of all, in central Cusco.[41] Concurrently, under the Sapa Inca Pachacuti, this small city-state was promptly and comprehensively rebuilt into both an imperial and a spiritual hub. Cusco, or rather Qusqu (in Quechua), embraced the true meaning of its name: the navel of the world.[42]

In combining the rationality of a grid system with the form of a puma – the animal the Inca believed symbolised the power of the earthly realm, *kay pacha* – the new city plan provided tangible expression of the Inca's distinctive understanding of geography and the universe.[43] As many cat lovers know, felines tend to think a lot about food when they're not having a snooze, and appropriately enough, Cusco's two most important sites were located in the privileged position of the puma's belly. First, the city's immense main plaza Hawkaypata and its *ushnu* not only served as the locus of political events and religious ceremonies, bringing together subjects from across the empire, but additionally constituted the crossroads where roads from all four cardinal directions, each associated with a different *suyu* and a different quarter of Cusco, met.[44] Consequently, Hawkaypata connected both logistically and symbolically the entirety of Tawantinsuyu, giving credence to the notion that in the Andes, all roads lead to Cusco.

A short walk away stood the other key structure – the Qorikancha – a temple dedicated to the sun god Inti, with walls coated in gold and a location carefully chosen so that it would be aligned

---

* 'Place where one is lost'.

with the sun on the solstices and equinoxes, enabling it to operate as a giant calendar.[45] As well as connecting all four corners of the empire and the earth with the sun, the Qorikancha stood at the intersection of around forty ritual pathways called *zeq'es*, which the Inca used to structure their empire's geography.[46] Though not necessarily straight or equal in length, the *zeq'es* were critical to the Inca's distinctive way of interacting with the world, as they informed the routes taken by pilgrims along or besides the Qhapaq Ñan, who in turn marked these Andean versions of ley lines with their shrines.[47] Four of these invisible lines accompanied the four main roads running from Cusco to each of Tawantinsuyu's quadrants as far as their physical geography allowed, while others pointed to the positions of different celestial bodies, and helped demarcate the boundaries of each Inca territory. With the hilltop fortress-temple complex of Saqsaywaman located at the puma's head, the rest of Cusco was outlined by its two rivers, meeting at Pumaq Chapan, the tip of the tail.[48]

The Inca obsession with civil and spiritual order wasn't limited to Cusco's design. For one, social stratification was forged into the urban landscape, elites dictating that only the highest-ranking people were permitted to live within the puma.[49] For another, and despite the empire's clearly defined social hierarchy, the Inca's favoured architectural style was strikingly consistent regardless of neighbourhood or an edifice's primary function. These groups of buildings surrounding a common courtyard, called *kanchas*, were typically designed in a rectangular or trapezoidal shape and constructed from fieldstones; although mortar wasn't uncommon, many edifices eschewed this material for precisely fitted interlocking stones, ideally suited to resisting all but the Andes' severest earthquakes.[50]

Other key settlements throughout Tawantinsuyu featured their own quirks, but in almost all cases the Inca demonstrated the same concern with building order through this conscious segregation of industries and classes, the provision of uniform structures and the construction of sacred sites, all connected by the Qhapaq Ñan. Quito was officially founded in 1534 by the Spanish, but had previously acted as the Inca's northern capital and is widely believed to be the continent's oldest continuously inhabited city.[51] Ingapirca, Ecuador's finest archaeological site, was first settled by the Indigenous Cañari people, but the Inca later absorbed it into their empire and constructed its Temple of the Sun, which, consistent with the Qorikancha temple in Cusco, simultaneously acted as an observatory.[52] In central Peru, Huánuco Pampa was, like Cusco, laid out around a huge rectangular plaza with an *ushnu* in the centre and pathways radiating out from this point, dividing the city into different parts.[53]

Ironically, an exception to the general rule that the Inca ensured their settlements were well connected by roads was the one best associated with this civilisation today. Situated deep in the Peruvian Andes and overgrown with vegetation, the citadel and probable royal estate of Machu Picchu – which appeared to have been built in the shape not of a puma, but of another sacred animal, the condor* – was never detected by the Spanish even as they ravaged much of the surrounding area.[54] In fact, as far as many foreigners were concerned, it was only 'rediscovered' in the early twentieth century by the American archaeologist and explorer Hiram Bingham.[55] The Inca Trail, now a world-famous hiking route across mountains and through cloud forests, essentially represented a side street to be traversed only on special occasions, its lack of regular use plausibly securing Machu Picchu's survival.[56] Another Inca city, Choquequirao, similarly appeared to escape foreign eyes until it was reached by the Spanish explorer Juan Arias Díaz in 1710. Three times the size of Machu Picchu, at

---

* In Inca belief, this represented a messenger connecting *hanan pacha* and Inti with the earth. The other animal in the trinity of sacred animals was the serpent, a representation of *ukhu pacha* and eternity, although the location of any snake-shaped settlement has yet to be confirmed.

its peak it was probably an administrative and agricultural hub, yet one whose remote location off the main northern route from Cusco has enabled it to remain remarkably well preserved ever since.[57] For the most part, however, the principal Inca cities were intentionally situated along the primary roads, which could reach widths comparable to a four-lane highway today.[58]

Notwithstanding the fact that the Inca constructed some major roads as well as perhaps the largest single-roofed structure in the Americas of its epoch (the *kallanka* or great hall at Incalla-jta, in modern-day central Bolivia),[59] as important as physical size was the notion that the Qhapaq Ñan allowed people to connect the human, natural and supernatural worlds. Building on an older pan-Andean belief that mountains are sacred, the Inca oversaw the emergence and transmission of a modified ideology that afforded the road system a paramount place. As well as construct-ing roads to reach new sacred sites on revered mountain summits, the Inca established their own pilgrimage routes, the most signifi-cant of which, given its direct relationship to the Inca creation myth, connected Cusco with Lake Titicaca. Here they venerated the remnants of former Tiwanaku structures,* worshipped at new temples dedicated to the sun and moon on the lake's islands, constructed a sanctuary on the Copacabana peninsula and built many *qollqas* to cater for pilgrims, all with the effect of creating a road-based pilgrimage system that validated their combined political authority and spiritual ideology.[60]

More broadly, the Inca period saw the expansion of a mountain-oriented practice that remains to this day: the creation and veneration of *wak'as*. Though the term is very general – in

---

* One structure that likely dates back to the Tiwanaku period now represents an inland Atlantis, an underwater temple only discovered by the modern world in 2000. Despite its unlikely geography today, evidence of a well-preserved, paved road, among thousands of artefacts, indicates that it was once greatly significant to Andean societies, including the Inca.

different contexts, it can refer to sacred phenomena in nature such as mountains and caves, or to sacred objects created or reworked by humans, such as piles of rocks (miniature mountains called *apachetas*) and sculpted boulders – *wak'as* were and still are closely related to the geography of the Qhapaq Ñan.[61] Two key points should be noted here. First, the roads were not merely convenient in allowing priests and pilgrims to meet for ceremonies, conduct rituals and make offerings; some were additionally designed to run close to the *zeq'es* radiating from Cusco, even if, on the return trip, the most dedicated would clamber up steep slopes and climb into ravines away from the road to follow the *zeq'e* more precisely.[62] Second, climbing mountains had previously been regarded as a brazen act of trespassing, for even though pre-Inca societies revered these natural features, they typically regarded them as the exclusive domain of *apu* mountain spirits. By avoiding punishment for ascending the Andes and subtly refashioning the stony landscape, the Inca were therefore able to quickly assert their authority over newly conquered populations, who agreed that these new leaders truly were the children of the sun.[63] Of course, not all rocks are *wak'as*, and any one sacred site can be interpreted in various ways. However, in the case of those laid or inscribed by people along the Qhapaq Ñan hoping to create and reproduce metaphysical nexuses between the mortal and divine worlds, they represent a distinctly spiritual version of graffiti, applying human meaning to some of the world's most formidable natural landscapes, while (intentionally or not) marking the presence of the Inca as a political entity.[64] Over time, even some of the tallest peaks have become adorned with *wak'as*, as travellers and highlanders seek to appease the *apus* and connect as directly as possible with the gods in *hanan pacha*, the upper realm.[65]

Inevitably, given Tawantinsuyu's scale and the short lifespan of the empire, Inca influence over their subjects' belief systems and geography was uneven. While localised Andean belief systems were increasingly integrated as a systematic ideology, different deities tended to be worshipped across the empire depending on local priorities: for instance, the earth mother Pachamama was primarily venerated by farmers, whereas the moon goddess Mama Killa was more relevant to fishers.[66] In general, inland areas, especially along the main highland artery linking contemporary Quito to La Paz, were affected more than coastal regions, where pre-existing rulers were typically given more autonomy over their affairs.[67] Nevertheless, in those regions criss-crossed most comprehensively by the Qhapaq Ñan, the natural world suddenly appeared less threatening. On lower slopes, the presence of *apachetas* came to symbolise a differentiation between human civilisation below and the wild, untamable peaks above.[68] On the main routes, numerous sacred sites lined the roads, proving that however forbidding the passage, others had managed to reach the other side. *Mojones* – piles of stones used as boundary markers and milestones throughout the Qhapaq Ñan (albeit placed according to the Inca's distance unit of choice, the *topo*, approximately 7 km/4 miles) – allowed travellers to ascertain their location in even the remotest areas.[69] And always conscious of the need to protect their precious crops, the Inca built order into the landscape by constructing walls alongside their roads in agricultural areas.[70] Suffice to say, desire paths, those trails of erosion created by shortcut seekers across so much of the world, to the chagrin of more intransigent planners, were unlikely to ever become a problem in Tawantinsuyu.

In almost every way conceivable, then, the Qhapaq Ñan was integral to the empire and to the profoundly regulated society three generations of Sapa Incas endeavoured to create. It was a communications system, allowing information to be transmitted to and from distant corners of a largely rural empire. It was a military thoroughfare, enabling troops to be moved quickly and efficiently across Tawantinsuyu. It was an instrument of the Inca's collectivist philosophy, facilitating the conveyance of workers and commodities, shared among places that had hitherto little to no interaction with one another. It was both a conduit of and a canvas for Inca spirituality, enabling the movement of pilgrims between newly connected places, and offering a locus for their sacred practices. And it was a proclamation of the Inca's dominion over lands across the western half of South America, which were now not merely connected to one another, but also reshaped according to a distinctly Inca understanding of the human, natural and spiritual worlds.

For around eighty years, the Qhapaq Ñan served its purpose for the Inca, as an indispensable instrument of implementing and bolstering their procedures and conventions across a large swathe of South America. Unfortunately for them, however, they soon found that building order through connection is a double-edged sword. The ambitious and pervasive road network they developed to acquire and administer new territories was fit for purpose only as long as they remained the dominant power. As soon as they were confronted with a more formidable adversary, the system they used to wield and reinforce their influence became a mechanism of their own downfall.

In the 1520s, the same decade as the Spanish and Portuguese found themselves at a diplomatic impasse over how they would divide their presumptive new territories in Asia, the Americas' greatest empire was thrown into turmoil with the death of Pachacuti's grandson and Tawantinsuyu's third Sapa Inca, Huayna Capac. The two events, occurring on opposite sides of the planet, were in fact related, originating in Christopher Columbus's first voyage to the Americas in 1492, a result of which was the establishment of an invisible boundary through the Atlantic demarcating 'Spanish' and 'Portuguese' territorial claims.[71] Over the following decades, Spanish conquistadors, seeking wealth and glory and serving both their country and the Catholic Church, brought hitherto unfamiliar American lands under the control of the world's fastest-growing empire, while spreading 'Old World' diseases to their unsuspecting communities.[72] Elite status in Tawantinsuyu provided no immunity to smallpox, measles or influenza, the former most likely being responsible for the deaths of Huayna Capac, his probable heir Ninan Cuyochi and much of the remaining Inca leadership.[73] Suddenly, this meticulously organised society was plunged into the chaos of civil war, fought between two half-brothers resolute that they were now the rightful leader of an empire already past its peak.[74]

Ever opportunistic, the conquistador Francisco Pizarro exploited both Inca instabilities and the infrastructure they had conveniently provided. In the first instance, in 1532 he captured the war's eventual victor, Atahualpa, in Cajamarca, an ancient settlement in northern Peru close to what is now one of the world's largest gold mines.[75] Not people famed for their integrity, the conquistadors, paranoid about an Inca counter-attack,* first took the low road by reneging on an agreement to release Atahualpa on the payment of an extraordinary gold and silver ransom, and then executed him despite leading him to believe that he would be spared, should he promptly convert to Catholicism. (The only accommodation they made was that they used strangulation instead of burning at the stake, their traditional sentence for heresy, but one horrific to a people who believed that this would prevent a person's soul from reaching the afterlife.)[76] From Cajamarca, the Spanish then chose the path of least resistance by travelling south along the Qhapaq Ñan to Cusco, slaying Andean challengers and seizing a variety and quantity of supplies beyond their wildest dreams from the *qollqas* along the way.[77]

As surprising as Tawantinsuyu's rapid implosion was, it was more than matched by the astonishment of the conquistadors on encountering the infrastructure developed by the Inca. Explorers and soldiers who were conditioned to believe that their nation was

---

* The conquistadors had reason to be suspicious – Atahualpa did order a secret attack. It just happened to have been targeted at his half-brother Huáscar instead.

superior were stunned to find a road system so well maintained and extensive that it eclipsed anything Europe had seen since the Roman period over a millennium before.[78] As far as the Spanish were concerned, only an empire – and a very powerful one at that – could have been capable of constructing the Qhapaq Ñan. Yet despite their amazement at the Qhapaq Ñan, the Spanish held minimal sentimentality for this and other Inca accomplishments, and instead set about refashioning them to better suit their own needs – and greed. No longer was Cusco the empire's crossroads, and the Qhapaq Ñan's eastern artery its main route. No longer were gold and silver coveted by the dominant power as symbols of both status (for only the royal family could wear them) and spirituality (being associated with the sun and moon, respectively).[79] Instead, the Spanish prioritised those roads that connected the regions richest in resources they desired as economic *commodities* – not least the extraordinary silver deposits of Cerro Rico, Potosí's veritable rich hill – with ports, including Lima, their new regional capital.[80]

A system conceived for a multitude of purposes and accordingly used to connect all manner of places was thereby transformed into a mere instrument of colonial extraction and trade,[81] a long-term legacy of which has been a general inclination among the independent Pacific nations to build road and rail routes that prioritise the coast. Almost any section that didn't serve this objective, including roads connecting Inca administrative, military, agricultural and ritual centres, was soon abandoned.[82] And all the while, stories of Inca wealth filtered back to Spain, attracting new generations of conquistadors expectant of finding the mythical El Dorado,* and accelerating Spain's continued modification of Andean geography.[83]

---

* Originally, this legend can be traced back to a ceremony practised at Lake Guatavita, close to modern Colombia's capital, Bogotá. Here, the ruler of the Chibcha people was covered head to toe in gold dust and floated out into the

With the former empire decimated by war, disease and the Spaniards' abuse-laden version of the *mit'a*, which saw many die working for a system repurposed to subsidise mining for private interests and the Spanish Crown,[84] it was inevitable that the Qhapaq Ñan quickly fell into disrepair. Alongside the replacement of a tribute system oriented towards public works by a tax-based economic regime that prioritised foreign coffers over local road maintenance,[85] the shoes worn by the Spanish pack animals of choice – horses and mules – tore up roads designed for use by llamas and alpacas, whose soft, padded feet imposed minimal impact on the road surface.[86] Roads that sharply ran uphill to give the same indigenous species access to upper pastures, in many places via staircases (llamas excel at climbing), were usually disavowed by the Spanish, who struggled both with altitude sickness and with the Inca decision to situate many of their roads high in the mountains.[87] Over time, and reflecting their very different engagement with Andean geography, the Spanish reconstructed choice sections of the road system, building thoroughfares capable of accommodating wagons, and either replacing their detested, dipping rope bridges with stone arches (such as that near Abancay in Peru, which can still be seen today) or stabilising them with metal cables to provide a flat walkway.[88]

Throughout the former Tawantinsuyu, Inca settlements and landscapes were similarly revamped to serve and gratify the new occupiers in town. In Cusco, the Hawkaypata was replaced by a

---

lake, where he would throw offerings of gold wares into the water and finally jump in himself, emerging clean. On learning this story, the conquistadors, most of whom had been raised in rural poverty and were interested in becoming improbably rich, quickly sought out a gilded man so well heeled he could discard precious metals into a lake. They found the lake – and initiated one of several missions to extract the gold at the bottom – but they could not find their El Dorado. Over time, the Spaniards' vivid imaginations and, in many cases, illiteracy combined to concoct more fanciful tales of a golden city rather than a golden man, spread via hearsay.

smaller public square (now called the Plaza de Armas), and its surrounding *kanchas* substituted for churches and mansions.[89] The gold of the nearby Qorikancha was promptly looted, while the stone of its temple was used to build new structures such as the grand church and convent of Santo Domingo directly atop its foundations.[90] (Even so, perhaps the Spanish should have placated the *apus* like their Inca counterparts, for in a clear testament to the quality of the latter's craftsmanship, the remaining walls of their temple are still intact, while the church has suffered severe earthquake damage over the centuries.)[91] In Raqchi, the tallest known Inca structure, a temple dedicated to the creator god Viracocha, was almost entirely demolished and replaced with a Catholic church.[92] Further south, another church was constructed using stones from Tiwanaku at the high-altitude crossroads of Laja – a site that could have become La Paz, had the city's founder not been deterred by two freezing, sleepless nights and decided to establish Bolivia's future de facto capital in a nearby bowl-like depression instead.[93] And all along the Qhapaq Ñan, the sacred sites created by Indigenous communities were intentionally vandalised, whether through the painting of Jesus on *wak'a*s, or adorning *apachetas* with Catholic crosses.[94]

This strategy of desecration and destruction, which allowed the Spanish to introduce an alternative imperial-religious ideology to their new territories, was soon accompanied by a more pernicious variant of the Inca's forced resettlement of recently conquered populations. Called *reducciones* ('reductions'), these new settlements were founded throughout the Spanish Empire to coercively assimilate Indigenous people to Spanish norms and evangelise them into the Catholic Church.[95] In the Andes, more than a million people were suddenly compelled to adjust to living in hundreds of oversized model towns designed according to a foreign template, their largely rural and periodically transient way of life supplanted by an increasingly urban, static and draconian existence.[96] With the joint goals of 'civilising' Indigenous populations and enhancing colonial surveillance and discipline, *reducciones* were consistently planned so that a central square would be flanked by buildings reflective of a new kind of order, both religious (a church) and punitive (a jail). Meanwhile the relatively few roads still maintained by a dwindling number of workers were reformulated to enable travel between *reducciones* and Spanish-controlled political centres instead, and hillside terraces, for centuries cultivated according to their specific microclimates, were amalgamated, rendering them prone to water erosion and infertility.[97] Quickly, the Andes thus came to embody a more European understanding of space. The Qhapaq Ñan, once a public good for the entirety of Tawantinsuyu, became a relic of an empire already wiped from the map.

Time has not been kind to the Qhapaq Ñan. Conflicts and landslides have left sections abandoned and uncared for.[98] In some areas, the stones that once paved the roads have been repurposed to build new houses, farms and fences. Though certain sections have been transformed into modern roads – including some of the contemporary Pan-American Highway – the majority of the system has fallen into disuse, evidence of its grand history reduced to overgrown trails where the earth appears to have once been compacted and elevated, and short stretches of stone paving.[99] As archaeological sites, roads are unusual in that they typically decline through too little use, rather than too much.[100] While this isn't to recommend that hordes of tourists stomp the remaining sections of the Qhapaq Ñan back into relevance, it is encouraging that in 2014, the former road system and 273 of its component sites were inscribed on the UNESCO World Heritage List. Originally championed by Ricardo 'El Caminante' Espinosa, a Peruvian trekker who walked and documented thousands of miles of the system over the course of several years, the successful joint application between all six relevant countries has helped increase collaboration and regional integration between nations whose modern histories have not always been harmonious.[101] In the process, it has brought new attention to an infrastructure that – despite its sheer size, and despite its centrality to Indigenous cultural traditions and a distinctly Andean worldview – has lain hidden, on the planet's surface, yet beyond most people's gaze. Now, careful management, following in the Inca legacy, is necessary to preserve both the remainder of the Qhapaq Ñan, and the diverse tangible and intangible cultural heritage along the way. Peru's limited provision of permits to trek the Inca Trail to Machu Picchu already demonstrates that the descendants of Tawantinsuyu continue to be attentive to the relationship between geographical connection and order.

Through the Qhapaq Ñan, the Inca wrote their history onto the Andean landscape, and either initiated or popularised traditions that remain to this day. Some modern pilgrims still recreate the *zeq'es* through carefully arranged itineraries and rituals, while in certain regions, travellers retain the practice of building roadside *apachetas* to ask for safe passage.[102] Spilling a small amount of *chicha* on the ground for the Earth mother Pachamama before drinking the rest continues to be practised in some regions, even if the majority of people now fuse Indigenous beliefs with Catholicism.[103] Beyond the best-preserved sites and sections of roads, most of which tend to be in drier, isolated areas such as deserts and non-arable plateaus,[104] Inca geography can be observed in the specifically sacred name of one of Cusco's central streets, Calle Pumacurco. More subtly, it is significant that living close to the former Qhapaq Ñan continues to be associated with superior educational attainment, nutrition and wages,[105] and it is no coincidence that vendors today will often seek roadside sites where they can most easily sell their handicrafts, evidence that geographical connection can present opportunities to those best positioned to make use of them.* The road system may no longer

---

* Nor has the Qhapaq Ñan's importance remained limited to the Andes.

be conspicuous, but it has not been erased from the Andean palimpsest, either. With careful preservation, this beating heart of the region may even be strengthened again.

Five hundred years of general disuse have reduced the lifeline of Tawantinsuyu to a dispersed monument. However, as a symbol of humans' ability to connect the most unconnectable places, and to work not against nature but with it, the Qhapaq Ñan has few peers. The Inca's greatest construction project in turn built an empire, and gave deeper meaning to Pachacuti's personally chosen moniker, 'Earth Shaker'.[106] The Qhapaq Ñan allowed the Inca to establish control over the means of production, the exchange of products, the administration of people and the practice of religion, undaunted by terrain, climate and the natural hazards these can pose. Whether by mobilising their troops, assigning diplomats to coveted territories, relocating potential insurgents, creating dependency through the imposition of a tribute system, removing bridges to constrain enemy movements or persuading the inhabitants of newly annexed lands that Tawantinsuyu's borders marked the expanding frontiers of an enlightened society, the Qhapaq Ñan allowed the Inca to exert power in whatever way would best accomplish their goals. Barren regions were converted into arable and pastured lands and connected to agricultural centres;

Just as the Inca used their roads to transport wool (enabling the production of traditional ponchos), cotton, corn, potatoes, tomatoes, chilli peppers, peanuts, quinoa, coca (a leaf traditionally used for medicinal purposes, but now more widely associated with the stimulant drug cocaine), tobacco, amaranth, gold, silver and more, over the following centuries, all these products were spread in one way or another throughout the world, changing cultures and lives in the process. At least two words in English evoke a distinctly Andean history: guano, a dung fertiliser transported by the Qhapaq Ñan from the coast to the highlands; and jerky, which Quechua speakers call *ch'arki*, a preserved meat product to be carried on long walks. The Andean practice of freeze-drying potatoes for travel purposes also long pre-dates the contemporary equivalent of this dehydration process.

precipitous mountains, sacred to diverse populations, were made accessible from temperate plains; a cornucopia of produce from disparate ecosystems was transported to wherever it was needed or desired. As much as the conquistadors believed they were civilising a region they considered peculiar, the forward-thinking Inca and their predecessors had already utilised the talents of people from all walks of life to operate in some of the planet's harshest environments, and to integrate even the most fragmented places. From rest stops to road signs and from maintenance crews to retaining walls, the pre-industrial world's most advanced road system unwittingly anticipated many of the features of our present-day transport infrastructure, all while keeping users corporeally connected to the ground beneath their feet.

2

Extraction: Mozambique's railways

---

*The main fight of FRELIMO is
against human exploitation.*

Samora Machel[1]

	International border	
	Railway	
	River	
■	Settlement	

TANZANIA

ZAMBIA

MALAWI

Lake Malawi

Lichinga

Pemba

Nacala line

Nacala

Cuamba

Nampula

Nkaya

Zambezi River

Moatize

Blantyre

Tete

MOZAMBIQUE

Sena line

Tete line

Sena

ZIMBABWE

Harare

Quelimane

Machipanda line

Machipanda

Mutare

Trans-Zambezi line

Dondo

Beira

Bulawayo

Indian Ocean

SOUTH AFRICA

Limpopo line

Inhambane

Xai-Xai

Lourenço
Marques line

Ressano
Garcia

Pretoria

Matola
Maputo

Matsapha

Johannesburg

ESWATINI

Goba line

Angola

D.R.C.

Tanzania

Malawi

Namibia

Zambia

Zimbabwe

Botswana

Mozambique

Madagascar

Lesotho

Eswatini

South
Africa

Ferroequinologists of the world, your attention please! No longer can trains be lazily derided as musty capsules with tired seats, and trainspotters typecast as loners in raincoats scribbling notes under a gloomy sky. Have we forgotten the magnetism of the Orient Express, or the mystique of the Trans-Siberian? Today, the rusty blood vessels of the industrial world are undergoing a major reanimation. While the oil-rich United Arab Emirates plans to extend its freight network to passengers and create a rail connection with Oman, China proceeds unceasingly in developing the world's most expansive high-speed rail network, where travellers can now ride from Beijing to Shanghai – a distance similar to that between New York City and Chicago – in under four and a half hours. India is on the cusp of offering a high-speed service of its own, starting from the same city as its original passenger line in 1853 but travelling fifteen times further, from Mumbai to Ahmedabad rather than Thane. Not to be outdone, Japan, a country long popular among railway buffs, is constructing its first magnetic levitation (maglev) Shinkansen,* whose breakneck 505 km/h (314 mph) speeds will allow passengers to travel the 350-km (215-mile) journey between Tokyo and Nagoya in just forty minutes.

Recognising the growing demand for green, affordable ex-periences, a new sleeper service aims to adjust Europeans' penchant for short-haul city breaks by reinvigorating night-time rail travel, connecting Brussels with Prague and in time Barcelona

---

*Bullet train, a name rather more evocative than the literal Japanese meaning: 'new trunk line'.

and Stockholm as well. Meanwhile, in Estonia, Latvia and Lithuania, an ongoing high-speed project seeks not merely to connect these states physically with Poland and Finland, but by replacing their maligned Soviet-era, eastward-looking infrastructure with a direct link to their closest friends, additionally promises to reintegrate them symbolically within Europe. Even the United States, a country long criticised by rail aficionados for prioritising motorists over public transport users, is now constructing a high-speed line between California's Bay Area and Greater Los Angeles, and a separate route from LA to Las Vegas, in an effort to reduce carbon emissions and stimulate economic growth. Along with new routes in countries as far afield as Mexico (the tourism-boosting Tren Maya) and Indonesia (whose high-speed line between Jakarta and Bandung is fittingly called Whoosh*), trains are finally reclaiming some of their former glory and glamour.

Admittedly, railways connote very different things in different places. The trains in a corner of south-eastern Africa would never be considered swanky, and are unlikely to be used by more than the most adventurous – and patient – rail enthusiasts in the near future. However, despite their lack of international renown, these human-made imprints on the land hold a particular significance regionally, containing within them stories from the past whose legacy continues to shape this area's present and future. Nowhere is this more apparent than in Mozambique, a country larger than any in Europe (bar Russia), whose surprisingly sparse and disjointed railway network is in no way commensurate with its size and shape. For a century and a half here, different powers both internal and external have recognised the possibilities railway infrastructure affords for extracting value and moulding places according to their interests, rendering Mozambique's railways

---

*An acronym of the Indonesian *Waktu Hemat, Operasi Optimal, Sistem Hebat* ('Saving time, optimal operation, great system'), although the onomatopoeia has the advantage of needing no translation.

something of a proxy indicator for geopolitical issues stretching far beyond the country's borders. Even now, Mozambique's unusually patchy rail system holds a specific allure for foreign enterprises, some of which choose to strategically extend those sections that promise advantages for themselves, but much less so Mozambicans. As a consequence, new efforts to expand Mozambique's railway infrastructure inevitably raise tough questions as to the true purpose of investing in this often neglected part of sub-Saharan Africa. Will Mozambique one day host a rail network that meets its own needs? Or are the updated lines simply more of the same, a thinly veiled attempt to subjugate Mozambique and its people to the shifting ideological priorities of international powers, while exploiting its natural potential in the process?

These questions are worth asking, for since the development of the world's first railway in England in 1825, few countries have played host to as wide a range of political stakeholders and been at the mercy of as many conflicting interests as Mozambique. As late as the 1890s, select portions of the country were ruled by the Gaza Empire, building on a far more extensive history of kingdoms such as the Maravi (which gives neighbouring Malawi its name) and the Mutapa. Prior to becoming a key focus of Europeans' 'Scramble for Africa' in the 1880s, Mozambique had already played host to foreign missionaries and traders for centuries, a legacy memorialised in the name of both the country (likely a derivation of the Arab slave trader Mussa al-Bik)[2] and, formerly, its capital (Maputo, previously called Lourenço Marques after a Portuguese explorer). Even Mozambique's independence in 1975 following a bloody war with Portugal failed to provide any sort of respite from competing forces and ideologies, as the country became one of the most brutalised and yet widely overlooked arenas of the Cold War. Bringing together African nationalism, Marxism–Leninism and neoliberalism, while bearing some of the brunt of an apartheid government, Mozambique's modern history has been eventful, to say the least.

But amid this flux, there is one important area of consistency: railways. Seemingly mundane on the surface, Mozambique's railways remind us that building connections into our planet's geography is rarely innocuous. Rather, these actions can be laden with power, oriented towards exploiting environments and fellow people, and can recraft the landscape in such a way that it advantages only the dominant. Constructed by some and destroyed by others – and accordingly containing both stories of a fractious past and seeds of a fragmented future – Mozambique's railways have persistently found themselves at the heart of the region's history.

To understand and appreciate why Mozambique's railways are so noteworthy, we must start by comparing the country's shape and the distribution of its cities with its rail routes. Whereas most other national rail networks discernibly connect a country's principal settlements in a relatively logical manner, Mozambique's three biggest cities – Maputo (plus its largest suburb Matola), Beira and Nampula – are conspicuously *dis*connected. Mozambique is approximately 2,000 km (1,250 miles) long from north to south, so it might have seemed sensible to provide some sort of inland transport infrastructure to link its furthest extremities. Yet most of the existing routes run more or less parallel to one another across the far shorter distance from the country's western borders to the coast. This is not due to any limitation imposed by physical geography; far from it. The main mountain ranges in Mozambique run north to south, forcing certain east-to-west lines to encounter rather than evade these formidable obstacles. In terms of ease of construction, the area of Mozambique best suited for a railway would probably be north to south along the less marshy sections of the low coastal plains. Yet no such railway is to be found. As perplexing as this situation may appear to someone who – quite logically – expects transport infrastructure to serve key population centres and communities scattered throughout a territory, while snaking around rather than through substantial barriers to movement wherever necessary, the reason is simple, and no less rational, once we consider whose motives are at play.

The railways were designed by colonisers, not to benefit Mozambican people but, in the most basic terms, to exploit them and their territory. In the first instance, resource extraction was the priority, the railways' role being to bring natural resources from the interior to the coast as efficiently as possible: it is no coincidence that the first Portuguese proposals for railways in Mozambique were made in the mid-1870s, immediately following the discovery of gold and diamond deposits in the Boer-run Transvaal Republic* over the border.[3] However, Portugal also had a broader geopolitical objective in mind, which transcended mere immediate economic gain. Since its imperial peak in the mid- to late sixteenth century, Portuguese influence in international affairs had declined precipitously, with the effect that two centuries later, it was on Europe's periphery not just geographically, but also with respect to the colonial order. Indeed, when Europe's great powers convened at the Berlin Conference of 1884–5 to regulate Africa's colonisation and partitioning, Portugal – the country that had spurred modern European interest in the continent in the first place – was dismayed and disturbed to find itself playing second fiddle to Germany, the United Kingdom, France and Belgium.[4] Nonetheless, for Portugal, the conference was also a wake-up call. Having already seen its dearest colony Brazil gain independence half a century earlier, the country realised that action was now necessary to ensure that 'its' African discoveries weren't relinquished to its European rivals.[5]

The most obvious solution to Portugal's colonial predicament

---

* Also known as the South African Republic, its borders are roughly the same as the contemporary South African provinces of Limpopo, Gauteng, Mpumalanga and North West combined.

was to develop new transport infrastructure in its two sizeable territories in Southern Africa: Mozambique and Angola. Not only would this strengthen Portugal's claims to sovereignty, but by connecting resource-rich regions to these routes, it would also prove an economic boon for an erstwhile maritime superpower now at a low ebb. A Portuguese-owned and -run railway would provide the means to transport Africa's mineral and agricultural wealth to its ports, and from there export them to major markets for wild profits – a possibility recently enhanced by the opening of the Suez Canal in 1869. No longer, Portugal reasoned, would it be dismissed as the poor man of Europe: its African colonies were about to fund and foment a new period of Portuguese supremacy.

Portugal had held an interest in Mozambique since the late fifteenth century, so it might seem surprising that for almost 400 years, Portuguese colonists scarcely penetrated into Africa's heart. Though the authorities were well versed in strategies for manifesting control of land – in particular through leasing to settlers vast, feudalistic estates called *prazos*, most of which could be found along the Zambezi River[6] – few potential colonists wanted to permanently settle in remote areas where the climate was humid and where a host of diseases ranging from cholera to malaria were still endemic. The few Portuguese garrisons and trading posts to survive in the Mozambican interior tended to be those where settlers recognised that they might become marvellously wealthy through trading gold and ivory, as was the case of Tete, still the largest city on the Zambezi and one of the few major settlements in the country not located on the coast.

By contrast, the majority of Portuguese territory on the *continente escuro* ('dark continent'), as many called it, continued to be regarded as inhospitable and unknowable. Certainly, even though Portugal had claimed two Southern African colonies, Mozambique and Angola, which happened to be substantial, existing maps and commentaries consistently underestimated the sheer distance between their respective coasts,[7] suggesting that as far as the coloniser was concerned, most of the land was to be possessed but not valued, appreciated, surveyed or used. The Berlin Conference altered Portuguese priorities abruptly and entirely. Through its principle of 'effective occupation', the conference introduced an internationally recognised guideline for acquiring rights over territory, which required that powers provide evidence of colonial administration and infrastructure.[8] Whereas Portugal's relatively low-effort version of occupation thus far had resembled a colonial twist on finders keepers,[9] from 1885 it was suddenly expected to enter into treaties with local leaders and acquire concessions over many lands it had long presumed were its own. After a brief spell of introspection, Portugal soon assessed the new precept as providing a potential opportunity: here was a last chance to revitalise and expand its once-great empire. All it needed to do was prove that it was, in fact, a competent colonial nation. And so, driven by jingoistic fervour and a renewed vision of African development, Portugal committed itself to a railway-building programme.

The reformulated empire Portugal envisioned necessarily differed from that of its previous international apex, which had been founded on the allure of discovery and evangelisation. In place of 'brave' explorers and 'virtuous' missionaries sailing the high seas on the joint behalf of nation and God, the new focus was more banal, bureaucratic and worldly. It was also, in theory at least, easier to realise. Through executing and promoting African railway projects, engineers such as Joaquim José Machado and members of the new Sociedade de Geografia de Lisboa (Lisbon Geographical Society) saw a possible avenue towards their induction in the pantheon of Portuguese imperial icons, among the likes of Vasco da Gama, Ferdinand Magellan and António de Andrade from centuries before. All of a sudden, civil service in Portugal's Southern African territories was a source of phenomenal prestige, as young individuals could acquire unrivalled technological and administrative training in lands many deemed key to Portugal's future.[10] Having been largely neglected for decades, Mozambique and Angola started to be described as rich and fertile, capable of sustaining viable mining and agricultural economies close to South Africa's rapidly growing markets. Eventually, many colonists expected, the Portuguese territories would become important consumers themselves, with the requisite wealth to purchase Portuguese goods and thereby strengthen the colonial machine. While others were rather more realistic about Mozambique's economic potential, rightly recognising that the greatest mineral wealth was actually over the border and that this territory was not, in truth, exceptionally fertile, most chose to overlook such concerns and instead mythologised it as a pre-eminent stop on da Gama's travels to India, at a time when Portugal – not those dastardly Brits – ruled the waves.[11]

Through imposing their hard infrastructure on lands that for centuries had typically been used as pastures or hunting grounds, the Portuguese colonists sought to create a new economic hierarchy in Mozambique not dissimilar to the Spanish in the Andes before. Railways and harbours as well as roads would provide access to select sites – mines, plantations and trading posts – where profit-making was the goal. Capitalist concepts of 'resource' and 'commodity' were used as justification for manipulating the landscape,[12] compelling Indigenous African people to adapt and assimilate, albeit without any of the privileges available to their white counterparts.[13] Over time, the Portuguese hoped, their railways would not only make their presence in Southern Africa more visible, but would also render foreign colonial powers dependent on them. After all, no other transport infrastructure was as capable of moving labourers and goods across expansive swathes of territory, of particular benefit to Mozambique's mineral-rich but landlocked neighbours Rhodesia, Nyasaland, Swaziland* and the Transvaal.[14]

Railway development hence combined pragmatism with

---

* Respectively now Zambia and Zimbabwe, Malawi and Eswatini.

idealism, practicality with symbolism, and avaricious intent with mythological conviction. By offering the possibility of 'taming' Mozambique's rocky uplands and insect-infested swamps – and, with clear racist overtones, 'civilising' its unsophisticated people – railways were regarded as vehicles by which Portugal would reassert itself at the top table of global powers.[15] By imbuing certain sites with a distinctly 'European' grandeur – the best example being Lourenço Marques's Beaux Arts masterpiece,* for over a century a strong contender to being the most beautiful station in the world – the Portuguese strove to ensure that the users of their lines would recognise their merit as a commanding colonial nation.[16] To amplify its message, Portugal aspired to connect the Atlantic and Indian oceans using this infrastructure, recognising that Mozambique and Angola conveniently sat on opposite coasts – a goal even stated in the name of one of its railway enterprises, Companhia Real dos Caminhos de Ferro Através de África (Royal Railway Company through Africa).[17] As a consequence, a railway between them would enable Portugal to challenge Britain's claims in the African interior (including the majority of present-day Zambia, Zimbabwe and Malawi) and ultimately to assert sovereignty over a continuous west–east land corridor across the continent. In this sense, exploitation pertained not merely to natural resources, but to the distinctive configuration of the region's geography as well.

To ensure any European competitors were taking notice in the aftermath of the Berlin Conference, in 1885 Portugal published a provocative *mapa de cor rosa* – the 'Pink Map' – to represent its interconnected territorial claims in Southern Africa,[18] using a similar tint to that popularised by the British. Though the United

---

* Characterised by an elaborate facade, grand arch and copper central dome, this station, which replaced the basic original opened two decades earlier, was designed in part to eclipse its rival, Johannesburg in modern-day South Africa.

Kingdom refuted many of Portugal's claims of effective occupation in Africa's interior and, in 1890, duly issued an ultimatum demanding the prompt withdrawal of Portuguese forces, it was clear that the message had been successfully received: Portugal merited serious diplomatic engagement once again.[19] If railways could constitute at least a starting point in future negotiations with the great powers, maybe, just maybe, Portugal would be back on track.

Over the following decades, subsequent maps would attempt to convey a sense of Portuguese greatness – and confidence – to European audiences, although none was more evocative than that produced by Henrique Galvão for the Porto world's fair of 1934. By superimposing all of Portugal's remaining colonies onto Europe, this map showed that Mozambique and Angola alone could together cover most of the European mainland, stretching from central Spain to Russia's western borders. Still, contained within the map's somewhat defensive title, *Portugal não é um país pequeno* ('Portugal is not a small country'), was an uncomfortable truth. As much as Portugal wanted to prove to the world that it was still a global power, its efforts were tinged with an almost hysterical fear that foreigners would one day seize the territories it called its own.

Portugal's anxieties were far from unfounded. In fact, the Mozambican railway programme encountered significant obstacles from the get-go, with no country vexing the Portuguese colonial mission more than the most avid railway developer of the age, the United Kingdom. Following the vision of Cecil Rhodes,* a mining magnate who served as prime minister of the Cape Colony (essentially the western half of today's South Africa) from 1890 to 1896, the

---

* A highly controversial figure owing to his passing of various racist laws – which would later help inspire South Africa's apartheid system – and his desire to create a secret society capable of bringing about the United Kingdom's colonisation of the world.

British were already active in using railways as tools of imperialism, capable both of transporting valuable minerals from mines and of 'domesticating' underdeveloped lands.[20] Through his British South Africa Company, a chartered association with paramilitary functions which effectively operated as an arm of the British state, Rhodes had already overseen the development of substantial infrastructure in Southern Africa, and viewed railways as essential to his own ambitious connective vision. Just as the Portuguese dreamed of linking Africa's west and east coasts, Rhodes ultimately aspired to create a north–south railway along a continuous 'red line' of British territories stretching from Cairo to Cape Town, to solidify the United Kingdom's control over trade on the continent, unify (and justify) its colonial possessions, and enable the rapid movement of troops in response to any military threat.[21] Though this project was never realised in full* due to a combination of Germany's own objectives of creating a west–east land belt called Mittelafrika ('Middle Africa'),† and the significant geographic, climatic and economic challenges of building a 10,000-km (6,000-mile)-long railway, it is easy to see why the British would be unsympathetic to a rival European colonial nation on turf they coveted.

---

* Cecil Rhodes would probably be rolling in his grave if he knew that a Cairo–Cape Town transport link – a highway, to be specific – has since been developed not by the British, but by the African people he doubtless would have maligned.

† An interesting border legacy of European geostrategy in Africa is the Caprivi Strip, a 450-km (280-mile)-long panhandle protruding like a bony finger from north-eastern Namibia to the Zambezi River. Instead of belonging to Angola, Zambia or Botswana, all of which share lengthy borders with it, the Strip was acquired by Namibia's predecessor, the colony of German South West Africa, to give German colonists direct access to a navigable water route to their East African territory. Or so they thought – for just a short distance east of the Caprivi Strip along the Zambezi River is Victoria Falls. At over 100 metres (350 feet) tall, it is not exactly a practical conduit to the Indian Ocean.

The Lourenço Marques or Delagoa Bay line, the country's proud first foray into African railway development, epitomised Portugal's struggles to wrest itself from British authority in the region. Back in 1877, the United Kingdom had sabotaged a similar project between the Transvaal's gold fields and the Mozambican port capital, suspecting, rightly, that the intent was to allow the Portuguese to bypass and compete with the less convenient British harbours in Durban, East London, Port Elizabeth (now Gqeberha) and Cape Town, and to give the Boers a direct export route free from British influence. The Boers' achievement of self-government in 1881 should have re-energised the Portuguese plan, but the project was severely delayed by their American contractor's inability to meet deadlines or set fares low enough to undercut the British competition, opening the door for the British to intervene again. Even though the Lourenço Marques line did manage to open in 1890 between its namesake city and Ressano Garcia on the Transvaal border, and before long was extended as far as Pretoria, the onset of a new war between the Boers and the British before the century was over ultimately saw the United Kingdom strengthen its own control over much of a region it had no intention of vacating.[22]

Over the following three decades, the Lourenço Marques line was accompanied by further routes connecting the continental interior with the coast – the Machipanda, Sena, Goba, Limpopo, Trans-Zambezi and Nacala – alongside some minor routes serving short stretches of shoreline around Xai-Xai, Inhambane and Quelimane.[23] On the surface, these railway lines manifested Portuguese ascendancy in the region, and yet the reality was very different. As early as 1892, the same year Portugal was declared bankrupt (an event that would be repeated a decade later), Rhodes had actually mooted purchasing southern Mozambique and the Lourenço Marques line.[24] When this bid to bring Portuguese rail under British influence failed, the British adopted a more nefarious fallback, spreading rumours that Mozambique was a hot, rainy and diseased territory, and therefore less appropriate for conveying goods than

South Africa's railways and ports. As contingency plans go, it was effective, subverting the line's economic viability and attractiveness, exactly as planned. Not only were the Portuguese frequently impeded in hiring workers, welcoming new colonists and convincing exporters to send their shipments via Mozambique's railways and ports, they were also compelled to curry favour with the more powerful British instead of truly competing with them.[25]

As a consequence, Portugal was never able to develop a self-sustaining model of development in Mozambique. Whereas originally it had endeavoured to make foreign powers dependent on its colonies, the relationship actually worked in reverse. On occasion, the Lourenço Marques line did at least enable a degree of symbiosis – in 1909, South African gold mines in the Witwatersrand were granted exclusive rights to recruit cheap Mozambican labour as long as they sent a fixed and significant amount of their export traffic to Lourenço Marques's harbour, benefiting both colonial powers economically while reducing African men to mere commodities* – but in the main, Portugal remained subservient to British interests.[26] Most disquieting of all from a Portuguese perspective, the country could neither afford nor operate the infrastructure that had quickly become the lynchpin of the Mozambican economy, instead requiring external investment and management. And predictably, inadequate investment in Mozambique's railways left long sections vulnerable to rapid decay. Almost immediately, a combination of shoddy workmanship and heavy rains led to heavy damage on the Lourenço

---

* Portugal had prohibited its nationals from engaging in the slave trade in its territories in 1836, but the authorities in Mozambique typically turned a blind eye to continued exports of enslaved people not only by Portuguese but also by African, Arab and French traders. The Mozambican labourers employed in South African gold mines were not technically enslaved, but nor were they free. Often locked up so that they wouldn't desert, many died of tuberculosis or silicosis while travelling on railway carriages designed for freight or animals.

Marques line near the Transvaal border. Within a matter of years, this same line was regarded as too fragile to handle its increasingly heavy traffic, necessitating the development of a new route connecting Mozambique with Swaziland, to be financed entirely by Portugal. However, this time the United Kingdom did not face any legal obligation to direct trade along the new line, weakening Portugal's resolve to extend the route across the Swaziland border while providing it with yet another unwelcome diplomatic dispute.[27] Taken in combination with more practical challenges, including the threat of contracting African trypanosomiasis (sleeping sickness) or malaria, and, in the case of the short Gaza line, Indigenous resistance to colonial occupation,[28] railway construction proved to be a more formidable challenge than all but the most realistic of Portuguese administrators had anticipated.*

Lacking in diplomatic nous, reliable finances or any coherent long-term vision for colonising a region whose sheer scale proved to be a mixed blessing, Portugal's purported development of Mozambique through rail connections thus remained piecemeal; its fragmented transport systems instead primarily advantaged the United Kingdom. In an effort to mimic the British but without the prerequisite economic and military capacity, Portugal established its own chartered companies in the 1890s and offered them concessions to freely exploit and administer large portions of Mozambican territory, their natural resources and their populations. In return, these companies would develop these regions' infrastructure, including transport as well as agriculture, social services and trade.[29] The problem for Portugal was that to construct railways, these companies could not rely on Portuguese funding and instead were compelled to seek out alternative sources

---

* Intending to transport cashew nuts from the Gaza Empire, the Portuguese chose to wage a military offensive against this region's king, Gungunhana, in 1895. With his capture and exile, Mozambique's long – albeit discontinuous – history of monarchy was brought to a close.

of investment, rendering them vulnerable to foreign control. In barely no time at all, the Companhia do Niassa in the north, the Companhia de Moçambique in the centre and the Companhia da Zambézia around the Zambezi River in between all ended up at the mercy of other nations.[30]

The development of the Machipanda line, one of the many that connected interior to coast, was a case in point. Unlike South Africa, which already had its own British-controlled rail infrastructure, the land to the north (later named Rhodesia after Cecil Rhodes) was lacking in transport connections to Indian Ocean harbours, and the closest options were in Mozambique. As a stipulation of an 1891 border treaty, Portugal was legally bound to connect this territory by rail with Beira, a port city it had recently established to prove its effective occupation of central Mozambique. Yet despite its obligatory origins, such a railway offered certain political and economic advantages to the Portuguese as well – as long as they remained in charge of the line. Fearful of the diplomatic ramifications of reneging on their agreement and thus appearing incapable as a coloniser, the Portuguese made the impulsive decision to hire a British agent to realise the project through its Companhia de Moçambique. It was a move they would later regret; the agent in turn passed his responsibility on to Rhodes' associates.[31] All of a sudden, the Portuguese were in charge in name only. Once the company's board of directors was purged of foreign representatives soon afterwards, British control was effectively complete.[32]

Such was the brutal, deliberate and occasionally underhanded efficiency of British colonial power, which would be repeated else-where in Mozambique over the following decades.* In a cashew

---

* In the far north, for instance, where the British not only prevented the Companhia da Zambézia from developing a Portuguese-run line towards Nyasaland, but also managed to persuade Nyasaland's governor to pay for a different route that advantaged the British.

nutshell, a railway-building programme that had promised to connect Portuguese-controlled territories together was instead allowing the British to hitch many of the same lands to their own growing empire. To make matters worse for the Portuguese, the more routes the British secured in Southern Africa, the better able they became to direct rail traffic as and where they desired. Quickly, the dilemma facing Mozambique's Portuguese colonial authorities became whether it was better to build Portuguese railways using British money than to have no railways at all. Believing – rightly – that a lack of action was tantamount to relinquishing its Southern African colonies, Portugal tried to strike a fine balance between welcoming British financial assistance, ideally from investors unaffiliated with the British government and its imperial agenda (as illusory as this tended to be), and putting itself at risk of grave and potentially expensive diplomatic and legal disputes over railway operations. However, over time the domination of nominally Portuguese companies by foreign men became a source of significant frustration and embarrassment for many Portuguese colonial leaders, even if few were as open as one president of the Companhia de Moçambique, Fontes Pereira de Melo Ganhado, who publicly acknowledged that his company had effectively become a department of Rhodes' British South Africa Company, instead of operating actual Portuguese railways.[33]

Gradually, then, Portugal forced itself further and further into a corner over railway development, being unable to afford the infrastructure it needed to recharge its economy and empire, but also increasingly resistant to letting go. With the United Kingdom and Germany planning for Mozambique's partitioning as early as 1898, Portugal realised that despite its financial crises, pursuit of its railway programme was critical to its future as a colonial nation, but in its determination to compete with these powers, all it did was increase its dependency on foreign support.[34] Further, even though these railways joined Mozambique to its neighbours and their mineral wealth, Portugal scarcely benefited, importing relatively little and instead operating more in the reprehensible export of African labourers.[35] Its foothold in Southern Africa perpetually precarious, instead of scrambling for new territories, Portugal was constantly left rummaging for partners and funds.

In late May 1926, the Portuguese mainland was plunged into turmoil of its own. The country, already destabilised by repeated coups d'état, saw nationalist revolutionaries under military command quickly overthrow their democratic but delicate republic and set in motion what would prove to be one of modern Europe's most enduring dictatorships. Looking on from afar, much of the Portuguese community in Mozambique welcomed António de Oliveira Salazar's *Estado Novo* (New State), believing that his defiant nationalist and pan-Lusophone* rhetoric would help reinvigorate the colony's languishing railway-building quest. They were left disappointed. Though Salazar's denial of future support to Mozambique's desultory chartered companies was welcomed by colonists hopeful that Lisbon was about to invest in new routes that truly served their interests, this never materialised.[36] Conspicuously, the two main routes to be built during his nearly forty-year premiership (the Tete line from Beira to the Moatize coalfields, and the extension of the Nacala line from the northern city of Nova Freixo† to Nkaya in Malawi) were both covered in foreign fingerprints.[37]

Still, the colonists' misgivings were nothing compared to those of Mozambique's Black population. Angered and perturbed by pervasive racism and violent repression, the Liberation Front of Mozambique (FRELIMO)‡ was born over the safer Tanganyika§ border in 1962, where it was armed and trained by anti-colonial

---

* Portuguese-speaking.
† Now Cuamba.
‡ In Portuguese, Frente de Libertação de Moçambique.
§ Now mainland Tanzania.

and anti-Western Cold War partners, including the Soviet Union and China.[38] Two years later, FRELIMO initiated a ferocious independence war that would change the function and functioning of Mozambique's railways for decades, as the connections built for territorial expansion and resource extraction suddenly became susceptible to assault.[39] In fact, the railways only became more vulnerable after Mozambique achieved its hard-fought independence in 1975, as FRELIMO suddenly found itself having to defend the same colonialism-tainted infrastructure from brutal guerrilla attacks by a new, anti-communist enemy.

By design, this Rhodesia-sponsored movement, the Mozambican National Resistance (RENAMO),[*] allowed Mozambique to descend into disarray. Most notorious for mutilating, raping and massacring rural residents,[40] RENAMO also strategically assaulted railway lines, bridges and rolling stock, bringing the new country's transport corridors to a practical standstill.[41] Sometimes RENAMO opted to ambush carriages and rob or kidnap their passengers. Other times, it used landmines to derail trains and destroy long sections of track – and just to add to the maelstrom, FRELIMO used the same weapons in defence.[42]

Already suffering from years of underinvestment and neglect, useful, connective infrastructure such as the Dona Ana – at one point one of the longest rail bridges in the world and an essential regional link to Malawi – was impaired to the point of being unusable. Further reflecting Mozambique's enduring potential as a link between interior and coast, when white minority-ruled Rhodesia was replaced with Black majority-governed Zimbabwe in 1980, RENAMO retaliated by destroying road, rail and oil infrastructure between the port of Beira and Zimbabwe's largest city near the border, Umtali (now Mutare).[43] At a time when the racist apartheid state was itself hamstrung by international sanctions and boycotts, South Africa strove

---

[*] In Portuguese, Resistência Nacional Moçambicana.

to sever Mozambique's connectivity with the outside world as well, reducing the recruitment of Mozambicans in its mines and curtailing the amount of South African rail freight passing to and from the port in Maputo.[44] More than merely a throwback to the colonial period, this 'transport diplomacy' policy helped disclose a fundamental truth about railways' value in the region. Although conceived as essential components of a potent, connection-based colonial ideology, railways were also manifestations of dependency: wherever one link broke, expansive portions of the region were prone to disintegrate. It is no accident that Zimbabwe and Malawi each came to deploy their troops in beaten-down Mozambican transport corridors on which they, too, relied.[45]

Frail in its own right, by the early 1980s, Mozambique's economy predictably collapsed. While collectivised agriculture had failed to stimulate the rural economy, the country's nationalised healthcare and education systems as well as its rail infrastructure were constantly undermined by RENAMO's attacks.[46] Intensifying FRELIMO's growing sense of crisis, the party was abandoned by its long-standing ally the Soviet Union, at a time when its ever more stringent Marxist–Leninist ideology drove most potential diplomatic partners, investors and donors to focus their attentions on other post-colonial African states. In dire need of an ideological makeover, Mozambique thus entered the next period of its mercurial evolution. On applying for membership of the decidedly un-socialist International Monetary Fund, World Bank and Lomé Convention,* financial and military aid began to pour in from nations that had previously disparaged FRELIMO's politics.[47] While RENAMO continued to be supported by its established anti-communist allies, notably South Africa but more covertly the United States as well,[48] FRELIMO

---

* A trade and aid agreement between a large number of African, Caribbean and Pacific countries and the European Economic Community (since rolled into the European Union).

forged new bonds of an increasingly commercial kind, including private investors interested in Mozambique's mineral and tourism potential. Although the post-socialist transition brought the pain of austerity, currency devaluation and rising costs, and Mozambique remained deeply fractured and disconnected by war, at least FRELIMO could no longer be viewed as a pariah by the capitalist world.[49] This reputation instead fell more and more on the country's south-western neighbour, whose relentless refusal to abolish apartheid, combined with its continued support for RENAMO even following a non-aggression pact in 1984, and its likely involvement in the plane crash that killed Mozambique's president Samora Machel in 1986, rendered it truly broken in international eyes.[50]

On shedding its Marxist–Leninist vanguard moniker in July 1989 and dropping 'People's' from the country's official name the following year, FRELIMO's renunciation of its socialist origins seemed complete. More importantly, and although their civil war continued into the new decade, it was clear by now that both FRELIMO and RENAMO were running out of steam. The conflict as well as drought and famine had depleted the energy of a population increasingly displaced across both the country and its borders. With little left to steal or destroy, RENAMO's purpose, if it ever truly had one, was gone. Following a partial ceasefire – notably, in the critical transport corridor connecting Mozambique with Zimbabwe – and over two years of negotiations, a peace agreement was finally reached in Rome in October 1992. Despite the war's unthinkable brutality, both sides were given a blanket amnesty in hopes of bringing together a profoundly disunited society. In return for free and fair elections, RENAMO agreed to desist from hostility and instead opted to become a legitimate political party. Reflective of Mozambique's enduring rifts and fragmented geography, FRELIMO's Joaquim Chissano won the country's first multi-party presidential election in 1994 thanks to his support base in the far north and south, while his RENAMO counterpart Afonso Dhlakama proved more popular in a continuous belt across the centre of the country, thereby dividing pro-Chissano territories into two.[51] However, with an impressive 87.9 per cent of the electorate voting in an election deemed free and fair by international observers (despite RENAMO's claims of irregularities),[52] Mozambique finally appeared to be ushering in a period of relative stability, in which railways, the scene of inconceivable carnage for over a quarter of a century, might instead become essential tools of a necessarily reconstructed landscape.

With the country deeply divided both politically and socially, Mozambique's combined post-colonial and post-socialist transition was bound to be demanding. Gradually, though, important steps are being made to unify a country that had never truly operated as one. Especially considering Mozambique's marshy terrain, poverty and sheer size, no accomplishment has been more impressive than the clearance in 2015 of every last landmine laid during the independence war and civil wars. Pioneered by the British charity HALO Trust, this initiative has allowed vast areas to be opened up for agriculture, and enabled formerly detached rural districts to be connected emotionally, if not yet practically. Although the gesture was more symbolic and political than directly geographical, it is also noteworthy that in 1995 Mozambique became the first country to join the British Commonwealth without having been a part of the British Empire, notwithstanding the efforts of Rhodes and his minions. This quirk, since emulated by Rwanda, Gabon and Togo, epitomised the country's growing aspiration (and, arguably, post-war desperation) to participate in international initiatives promoting peace and development both here and across every one of its borders,* not least with Nelson Mandela's

---

* The exception now is Zimbabwe, which was suspended from the Commonwealth under Robert Mugabe's controversial leadership in 2002, and which opted to withdraw altogether a year later.

post-apartheid South Africa. (Incidentally, on marrying Mandela in 1998, Machel's widow Graça attained the unique distinction of serving as first lady of two countries in the modern era.)

Railway lines had been crucial to the Portuguese strategy of reworking the landscape, bending it to suit their needs and desires. It should be no surprise, then, that among Mozambique's efforts to stitch itself together as well as better connect with its neighbours, the revival of said railway lines would play a key role. FRELIMO, Mozambique's dominant party ever since independence, continues to emphasise the need for a viable transport system to stimulate the country's economy, and, reflecting its about-face from its socialist past, is active in procuring contracts with private operators capable of rehabilitating the country's decaying infrastructure. The results of Mozambique's public–private collaborations have been rather impressive. With the support of investment and expertise from as far afield as Australia, Thailand and the United States – not to mention Mozambique's Lusophone partners Portugal and Brazil, and prominent railway enthusiasts India and China – lines both old and new are once again becoming operational in select portions of this sweeping nation. Additionally, the three most significant ports (Maputo, Beira and Nacala) have all been privatised and modernised, and are now capable of accommodating an unprecedentedly wide range of cargo.[53] After years of abuse, Mozambique's transport lifelines convey regular traffic again. Mozambique, ostensibly, has been reconnected. Nonetheless, an important question remains: to what end?

Striking in this regard is how Mozambique's railways still constitute three separate systems, each designed to export resources from the interior to the coast while ignoring one another and everywhere in between. Whereas Maputo in the south continues to provide a port for goods from South Africa, Eswatini and Zimbabwe, Nacala in the north does the same for materials from Malawi.[54] Encapsulating Mozambique's continued lack of internal connectivity, the centre system, whose port is Beira, reaches the borders of both Zimbabwe and Malawi and could therefore be joined to its northern and southern counterparts in theory, and yet it remains a discrete system. Considering the diversity of Mozambique's new partners, it may seem surprising that these mining companies, engineering firms, national banks and more are typically aligning their railways in much the same way as the Portuguese did in the colonial era. However, and partly owing to the World Bank's designation of financial support,[55] the preference so far has been to prioritise the country's existing tripartite transport corridors over the long-delayed creation of a single, cohesive transport network.

Though this distinctive configuration of railway lines is doubtless beneficial to the firms that use them to convey their goods, for passengers and smaller enterprises, it is far from fit for purpose. Consider the 700-km (450-mile) distance as the crow flies between Maputo and Beira. A railway passenger must travel first on the Limpopo line into the heart of Zimbabwe before heading east on the Machipanda line, effectively doubling the possible journey. Likewise, travelling from Beira to Nacala requires a connection through Malawi via the Sena and Nacala lines, rather than following the coast past Quelimane. The express international routes currently being planned also retain the colonial model, offering quicker connections between Mozambique and its neighbours[56] while neglecting the possibility of finally integrating the northern, central and southern parts of the country. Further, with every existing route being maddeningly slow,* lacking in services and therefore seats, and in most cases prone to derailments, there are regrettably few enticements to travel by rail, even though a theoretical demand does exist.[57] Relative to the total length of its main

---

* For example, Maputo to Ressano Garcia on the South African border is under 90 km (55 miles) but takes four hours, while Beira to Machipanda on the Zimbabwean border is around 300 km (200 miles) but involves an eleven-hour journey.

rail network – a good 2,500 km (1,500 miles) – Mozambique is, suffice to say, one of the least convenient countries in the world in which to travel by train.

It is also not as if users have many alternatives. Domestic air services remain minimal: many airports only offer flights on demand, and runways tend to be in poor condition. Heavy rains during the wet season regularly cause bridges to collapse and flood both rail lines and the country's mostly unasphalted roads, greatly limiting the transport options available at this time of year.[58] The few arterial roads that do exist largely duplicate the main rail routes, and are most useful to mining operators who want to avoid the prohibitive border fees occasionally demanded by Mozambique's neighbours. Those same vehicles, however, clog up the few, generally unpaved roads, and slash their lifespan.[59] With no road or rail running the length of the coastline, many communities along the Indian Ocean are essentially cut off from the interior, in some locations compelling residents to use dhows (small sailing vessels), which can travel in shallow coastal and riverine waters, to reach better-connected ports. And if an absence of human infrastructure doesn't pose an obstacle, then Mozambique's physical geography might: the country's sea ports require frequent dredging, limiting the use of large boats, while sandbars and fluctuating water levels have long constrained effective commercial shipping on even large rivers such as the Zambezi and the Shire.[60]

In a country that is also prone to disasters such as tropical storms, one may wonder why international companies are suddenly so interested in rehabilitating and building Mozambique's transport infrastructure. The answer, as ever, is minerals. Home to one of the world's most abundant and least developed reserves of premium hard coking coal, a resource particularly in demand for steel manufacture on the other side of the Indian Ocean, Tete province's Moatize region has become a mining mecca over the past two decades (though coal has been extracted intermittently by locals for nearly a century). Indeed, despite being situated in Mozambique's north-west and hence on the 'wrong' side of Malawi, Moatize is now seeing different companies, some of which operate as arms of their national governments, pledge to invest in railway construction in order to transport this coal efficiently to the coast. The problem is that circumstances can shift rapidly. As the Mozambican government has already seen to its cost, a foreign partner may fail to fulfil the conditions of a railway infrastructure project[61] or pull out of the country if it no longer considers it sufficiently profitable,[62] leaving those with the least say of all to pick up the pieces.[63]

Certainly, since its independence, Mozambique has been learning the hard way about the perils of engaging with international organisations and businesses. The International Monetary Fund's classic template of privatisation, free trade and minimal state funding has increased socioeconomic inequalities since independence, and without the barrier of protectionism, domestic industries – typically agriculture – tend to struggle to compete with their international rivals.[64] Building trust remains a challenge, a reality not helped by a 'hidden debt' scandal in 2013–14, revelations of severe corruption at the port of Maputo and widespread suspicion that government officials occasionally gloss over the decrepit condition of much of the country's transport system in order to attract private investors. More grievously, Mozambique's modern history of instability is still prone to rearing its ugly head, whether precipitated by RENAMO (which periodically disrupts Mozambique's roads and railways, and whose 2013 insurgency necessitated an updated peace agreement) or, increasingly, Al-Shabab, an Islamic State affiliate committed to waging a new campaign of terror against northern residents. In a country that has already undergone so much hardship, this group's predilection for executions, abductions and rape isn't just deterring much foreign financial assistance: it is also reawakening the agonies and traumas of a history most would prefer to leave behind.

Despite these disincentives, Mozambique's tremendous potential in agribusiness (cassava, sugar cane, maize, bananas, nuts, soybeans, among others), energy (natural gas, biofuels) and mining (not just coal, but diamonds as well) is destined to intrigue those foreign enterprises hopeful of beating the rush. Remarkable in this respect is how much the contemporary landscape of Mozambican railway politics resembles that of the colonial period. Sure, the parties involved in this oft-abused corner of Southern Africa are different, and it's true that even the most powerful multinational corporation in the world lacks the political control of Europe's colonial nations. However, in the sense that discrete companies are able to operate as proxies of foreign governments, dictate the course of railway lines connecting mineral reserves to ports and thereby render entire regions dependent on their effective functioning, the propensity among powerful outsiders to view Mozambique primarily through the lens of resource extraction is an enduring one.[65] Spotty though Mozambique's transport infrastructure undoubtedly is, it is never out of focus.

Simply put, for almost 150 years, railways have been built for foreign investors rather than residents. Over and over, Mozambique's geography has been manipulated to benefit the already powerful – at the expense of the people who live there. Railways may be just one part of Mozambique's modern, mechanised landscape, and a meagre one at that, but they are undoubtedly the most influential. Connection here has built not long-term opportunity, but enduring dependence. Whereas Mozambicans as forced labourers were long traded for traffic along rail lines designed for colonial profiteering and pride, now individuals have more autonomy in theory, but continue to rely on this transport to work in volatile industries run by faceless foreign figures. Yet somewhat paradoxically, *dis*connection has proved just as severe, for while Mozambique's relationships with the outside world have often made it a puppet of greater powers, its lack of linkages between north, centre and south can also cause it to splinter, each region more reliant on Mozambique's neighbours than on each other. One break in a transport corridor – damage to a railway line or economic crisis over the border – can cause the entire system to fail, paralysing a third of the country instantaneously.

Given these ongoing and unenviable challenges, Mozambique's future may appear desperate, but its past also offers some cautious reasons for hope. In the face of competition and oppression, of railways imposed on its still-fragmented landscape for the purpose of extraction rather than mutuality, time and again Mozambique has proved its ability to endure, to confront its troubles rather than deny their reality. Nowhere are its efforts to amalgamate its complex history and build solidarity from division more palpable than in its current national symbols. The Mozambican flag, without doubt one of the world's most distinctive, combines the colours of pan-Africanism, peace, revolution and natural resources with a miscellany of popular communist symbols, including a five-point star, a hoe and an open book, and, somewhat more dramatically, a Kalashnikov assault rifle and bayonet. Relatively speaking, its anthem, *Pátria Amada* ('Beloved Homeland'), might appear far more generic and innocuous, especially compared to its previous, unapologetic ode to the former one-party state, *Viva, Viva a FRELIMO*. However, on closer listening, hearing lyrics such as *Milhões de braços, uma só força* ('Millions of arms, only one force') and *Nenhum tirano nos irá escravizar* ('No tyrant will enslave us') belted out in full force, a distinctively Mozambican resilience comes to the fore. Here, Mozambique's demands are harder to miss than the piecemeal railway system scattered across its elongated territory. Poignant and persuasive, the message, if we choose to finally listen to it, is unambiguous: centuries of resisting oppression and exploitation aren't forgotten, and at long last, the country looks to sing with one voice. Maybe one day, further along the line, Mozambique's geography will be harnessed for Mozambique, and Mozambique alone.

3

Convenience: The Panama Canal

---

*No single great material work which remains to be
undertaken on this continent is of such consequence to
the American people as the building of a canal across
the Isthmus connecting North and South America.*

Theodore Roosevelt[1]

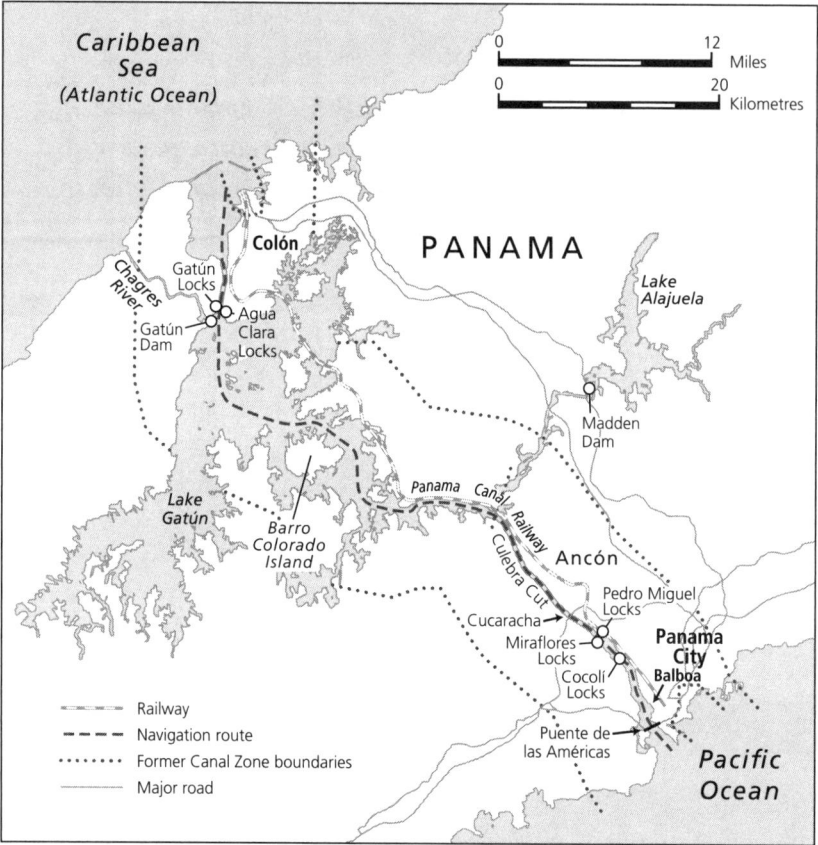

Caribbean
Sea
(Atlantic Ocean)

0   12
Miles
0   20
Kilometres

Colón

P A N A M A

Chagres River

Gatún
Locks

Gatún
Dam

Agua
Clara
Locks

*Lake
Alajuela*

*Lake
Gatún*

Panama Canal Railway

Madden
Dam

*Barro
Colorado
Island*

Culebra Cut

Ancón

Cucaracha

Pedro Miguel
Locks

Miraflores
Locks

**Panama
City**

Cocolí
Locks

Balboa

Puente de
las Américas

*Pacific
Ocean*

- — ⋅ — ⋅ — Railway
- – – – Navigation route
- ⋅⋅⋅⋅⋅⋅⋅ Former Canal Zone boundaries
- ——— Major road

Though far less famous than 11 November 1918, 9 November 1989 or 11 September 2001, 23 March 2021 lays claim to being one of the most globally significant dates in modern history. This was the day a Taiwanese-operated container ship caused global supply chains to seize up as they never had before. With strong winds blowing sand across the Suez Canal, the ship's giant hull suddenly deviated from its course, causing its bow and stern to become wedged on either bank. Immediately, hundreds of vessels found themselves stranded, foiled by the clogging of a waterway that typically sees more than $9 billion worth of goods and 12 per cent of world trade pass through daily.[2] Ranging from oil to grain and face masks to semiconductors, products earmarked for countless people saw inconceivable delays, until mercifully, a desperate team of dredgers and tugboats managed to free the *Ever Given* from its bondage six days later.

At a time when so much of the news cycle focuses on international barriers – on border disputes, trade wars, migration policy – this episode provided an important reminder that the vast majority of the world's population today lives in a genuinely interconnected society, where one breakage can have enormous, everyday ramifications for people all over the planet. We all know that a large proportion of our daily necessities comes from abroad, but until the *Ever Given* became stranded, we were seldom forced to consider how these necessities reach their destinations. How meaningful would a label identifying an item's country of origin – 'Made in China', 'Product of Brazil' – be if we weren't able to quickly and reliably transport that item from point A to point B? In recent years, the 'trade war' between the United States and

China has rightly garnered considerable attention for its impacts on millions of consumers' and employees' lives, including job losses, wage cuts and price hikes, but even beyond (generally momentary) diplomatic disputes such as this, we are constantly reliant on everywhere in the middle operating as expected. Indeed, as disconnected as we may feel from the places that produce and provide for us, we are even more likely to overlook the places 'in between', the links in the chain connecting the world together as a single unit and rendering the very idea of living in a fully self-sustaining nation virtually inconceivable.

With over 80 per cent of world trade volume still transported by sea,[3] a small number of key maritime corridors are particularly crucial to our daily existence, among them the aforementioned Suez Canal, the Strait of Hormuz between Iran and Oman, and the Malacca Strait between Sumatra and the Malay peninsula. Whenever a glitch occurs in the system – perhaps due to an earthquake or volcanic eruption or, indeed, a physical blockage such as a grounded boat – we all feel the results in rising costs and delayed deliveries. Moreover, because so many goods pass through the same, generally narrow stretches of water, it is no surprise that many such areas are magnets of disorder (not least the Gulf of Aden between Yemen, Djibouti and Somalia, whose notoriety for piracy has recently been worsened by numerous attacks on commercial vessels by Iran-backed Houthi Islamists), intensifying the risks of relying so heavily on just a few popular transport linkages.

The general tendency to ignore these intermediate spaces made the sudden surge in attention to the Suez Canal all the more remarkable. Whereas we usually satisfy ourselves with the knowledge that such conduits are simply 'there' and, like an electronic device before it breaks, take their smooth functioning entirely for granted, for six days in 2021, this waterway was as useful as a chocolate teapot. The boat blockage may not have been the most dramatic event in Suez's history – this is a place practically synonymous with a crisis in 1956, after all – but the image of a massive ship plugging a narrow body of water demanded the world's attention regardless. By contrast, when the planet's other most noteworthy waterway, the Panama Canal, was compromised by drought in 2023–4, few beyond the logistics industry and niche sections of the media seemed to be interested. For whatever reasons, 'lower than usual water levels' doesn't command quite the same degree of intrigue.

For over a century now, the Panama Canal has constituted one of the world's most important and fascinating technological accomplishments. As a physical connection between two oceans, constructed with absurd difficulty at the Americas' svelte waistline, it exemplifies our sheer bloody-mindedness when it comes to modifying the natural environment to better serve our interests. Offering convenience to consumers from Canada to Australia but bringing environmental complications locally, the canal has implications for Panama and beyond that continue to be as great as the feats of engineering behind its evolution. And yet the canal's connective significance isn't limited to its status as an infrastructural achievement without peer, a triumph over nature (or, more accurately, a unilateral revision to the balance we must find with nature) that reminds us that the planet's geography never need be static or inevitable.

As we have already seen, building geographical connections necessarily entails politics and clout, for nobody would choose to attempt a monumental undertaking simply for the sake of it, and as a country whose very independence is inextricably tied to the canal project, Panama has experienced more than its fair share of bureaucratic wrangling. Far more than being 'just' a link between east and west – and thanks to a quirk of Panama's physical geography, it really runs from north-west to south-east – the canal is in fact key to living in a globalised world. Well under half the length of its counterpart in Suez but four times as expensive and much more arduous to build,[4] the Panama Canal marked a paradigm shift when it was finally completed, cementing the

Americas as the planet's most internationally influential realm. Bringing together issues of national sovereignty and citizenship, geostrategy and diplomacy, all tied to the rise of a technologically advanced superpower, the Panama Canal is both a facilitator and an emblem of a distinctly US-oriented system of interconnectivity, whose relevance to modern society transcends logistics and trade.

Reflecting a time-honoured desire, if not ability, to manipulate parts of the planet to expedite maritime travel, both the Suez and Panama canals were dreamed up, in one form or another, long before they became realities. Whereas the ancient Egyptians constructed a series of small canals to connect, indirectly, the Mediterranean and Red seas via the River Nile nearly four millennia ago, a handy artificial waterway across Central America's narrow isthmus was proposed as early as 1529 by the Spanish explorer Álvaro de Saavedra Cerón. Though he died in the midst of elaborating his concept, Saavedra's maps of a new waterway, to be achieved by joining Panama's Chagres River with the Pacific as well as with its natural endpoint on the Caribbean, was popular enough in Spain that the king, Charles I (or Charles V of the Holy Roman Empire), commissioned geographical surveys to investigate this possibility. However, the challenge seemed too formidable,[5] and the idea was more or less shelved until the early nineteenth century, when the respected Prussian geographer Alexander von Humboldt tentatively suggested that a trans-isthmian canal might be built *somewhere*, just probably not Panama (he had overestimated the height of its mountains, and believed them to be insurmountable).[6]

Still, human idealism was not to be extinguished. Little more than half a century later, the two canal projects' fates became intertwined through the energy of the same French diplomat who formally presented the United States with the Statue of Liberty, in honour of the two countries' alliance during the American Revolution. Having overseen the successful construction of the

Suez Canal, which opened in 1869, Ferdinand de Lesseps enjoyed considerable esteem internationally and was naturally regarded as the right person to lead the creation of a second, more audacious canal project in Panama, then a Colombian province. As he reasoned, the scheme's potential was irresistible: not only would it produce a convenient shortcut for commercial traffic between the Atlantic and the Pacific, but as the Suez Canal was already proving, such an earth-moving enterprise also promised to make its shareholders filthy rich.[7]

Unfortunately, de Lesseps was only partially correct. Possessing neither engineering credentials nor a willingness to countenance alternative ideas, he stubbornly clung to the idea of building a canal *à niveau* – at sea level – as his Compagnie Universelle du Canal Maritime de Suez* (Universal Company of the Maritime Canal of Suez) had demonstrated so proficiently.[8] The problem was that Central America's steep mountains, dense jungle and coastal wetlands presented a far more diverse and forbidding terrain than Egypt's flat, sandy desert. Thus, despite the compelling protestations of Adolphe Godin de Lépinay, a fellow French noble whose genuine engineering expertise led him to conclude that a lock system would allow vessels to be lifted across the isthmus, de Lesseps learned the hard way that a Panama canal was far easier – and cheaper – to imagine than to actually build.[9] The bucket-ladder dredgers – mechanical vessels that use a continuous chain of buckets like a shovel to scoop up material and send it

---

* On 26 July 1956, Egyptian president Gamal Abdel Nasser selected de Lesseps' surname as the necessary codeword for his military personnel to seize this company's offices and nationalise what he regarded as an undesirable European colonial legacy on his country's soil. Later that year, the 10-metre (33-foot)-tall bronze statue of de Lesseps at the canal's entrance in Port Said was badly damaged and removed from its pedestal, in protest against France and the United Kingdom's support for Israel during the escalating crisis. To this day, the Egyptian public is divided over whether the (now restored) statue should be returned to its original position.

down a chute onto a smaller vessel sitting beside – that had been so effective in Egypt proved incapable of contending with Panama's thick mud and sticky clay.[10] To make matters worse, much of the spoil that was successfully dredged was deposited so close to excavation sites that as soon as heavy rains fell, it slid straight back into the incipient canal.[11] The French team had apparently not anticipated these periodic downpours, which characterise Panama as one of the world's wettest places. The regular deluges swiftly became the source of significant frustration and misery. During the rainy season, the Chagres became a ferocious torrent, capable of sweeping away many months of work diverting this erratic river.[12] Worse still, when the skies cleared, the pools of standing water left behind were particularly attractive to Panama's greatest killer, mosquitoes.[13] Alongside the risks posed by snakes, contaminated water and the construction project itself, the French team quickly found that they had greatly underestimated the task at hand,[14] prompting the famous cartoonist Thomas Nast to ask: 'Is M. de Lesseps a canal digger or a grave digger?'

Eventually, de Lesseps began to more seriously consider a lock-type canal, and in 1887, nearly seven years after construction began, he recruited one of France's leading engineers to design and build the locks. However, whereas he was immortalised by his concurrent, namesake project in Paris, Gustave Eiffel's decision to join the Panama Canal enterprise nearly ruined his life. With confidence declining among a French public no longer willing to invest in an endeavour that had once promised so much, little over a year later, de Lesseps' Compagnie Universelle du Canal Interocéanique (Universal Company of the Interoceanic Canal) was forced into bankruptcy. In one of the greatest scandals of the era, the Compagnie's directors were sentenced to prison time and fined for fraud and misuse of funds, and although both de Lesseps and Eiffel (who was only a contractor) were later acquitted on appeal, their personal dignity never recovered.[15] Eiffel withdrew from business and devoted the remaining three decades of his life to scientific research, especially in meteorology, aerodynamics and radio broadcasting, with his world-famous tower acting as a bespoke laboratory.[16] Notwithstanding a brief recovery as a *nouvelle* company in the mid-1890s, which willingly persevered with building a now in-vogue lake-and-lock system in spite of rapidly dwindling funds, the futile French venture was history before the century was out.[17] For the sake of scarcely 18 km (11 miles) of incomplete, disconnected canal, France had lost the lives of more than 22,000 mostly Caribbean workers, and poured away money more freely than the flow of the Chagres.[18]

In place of France, for decades one of the world's foremost industrial and colonial nations, an emerging international power was ready to step in. The United States had previously contemplated building a trans-isthmian canal of its own amid the California gold rush in the late 1840s, but a dispute with the United Kingdom saw the two sides pledge never to acquire exclusive control over any such waterway.[19] Still, their agreement did not dampen US interest in building such a canal, to the extent that in his very first address to Congress as president in 1869, Ulysses S. Grant called for its construction[20] and subsequently sent seven expeditions to study such a project's feasibility in Central America.[21] The resulting reports indicated that Nicaragua represented a more practical option than Panama (whose dangerous jungles Grant had experienced first-hand during his time in the army), Darién* or Tehuantepec, but Grant's administration was unable to come to terms on the prerequisite canal treaty before he left office.[22] Nevertheless, the idea continued to intrigue many in the United States, and gained new momentum with the Spanish–American War of 1898, when US naval forces doubtless would

---

* Ironically, the Darién Gap just 250 km (150 miles) east of the completed canal is now one of the world's least connected places, a lawless stretch of rainforest with no roads, severing the otherwise continuous Pan-American Highway network between Alaska and Tierra del Fuego.

have appreciated not needing to travel the hazardous two-month journey around South America.[23] With an outcome of this conflict being new territorial claims on either side of the Americas – including Puerto Rico and Cuba in the Caribbean, and Guam and the Philippines as well as recently annexed Hawaii in the Pacific – it was clear that no other country in the world stood to benefit as much as the United States from a direct water connection.[24]

Hoping to fortify their country's influence throughout the Americas, politicians and diplomats quickly set to work. In addition to nullifying its old treaty with the United Kingdom, which was now more concerned by war in Southern Africa and growing instability in Europe than with its land claims in the Americas, the United States would finally need to choose between Panama, where France had evidently struggled to build a canal, and another territory along the isthmus where the continental divide dips to even lower points, Nicaragua.[25] To ensure it didn't squander its newfound international reputation, not to mention the same sorts of money wasted by the French, the United States opted to establish a commission to assess the two possible routes.

But for a twist of fate – or more precisely, the pull of a trigger – Nicaragua and not Panama would have been indelibly associated with a globally significant canal. For decades, Nicaragua had been regarded as the frontrunner for any such project: being closer to the major ports of North America, the journey facing most ships would be much shorter, while the country's mammoth, eponymous lake already offered a convenient natural waterway across a good half of this territory. However, the assassination of President William McKinley catalysed a rapid shift in momentum. On succeeding McKinley, Theodore Roosevelt set about pressuring the commission to endorse his firm preference for Panama, which, he and his allies reasoned, offered numerous advantages of its own, both human and natural.[26]

For a start, the distance across the isthmus is much shorter in Panama than in Nicaragua, theoretically rendering the rest of the canal both quicker and, crucially, cheaper to build.[27] Further, the route would run through a single territory, whereas a canal cutting across Nicaragua's southern lowlands risked provoking a border dispute with Costa Rica. Unlike Nicaragua, Panama already offered decent infrastructure – notably, established harbours on either coast, connected by a railway built half a century earlier with US money to hasten travel to California's gold mines. By reviewing the Compagnie's failures, Americans could learn what not to do in Panama, while simultaneously capitalising on the fact it had managed to build certain segments already, including the navigable canals at the entrances.[28] It certainly didn't hurt that the French company, on learning that Nicaragua would be the likely choice, was now reportedly willing to sell its assets, including its equipment, maps, surveys, records and railway to the Americans for a bargain price.[29] Yet few factors were more compelling than Congress's growing fear before the key vote in 1902 that Nicaragua was simply too vulnerable to tectonic activity – a concern intensified by the eruption of 'the smoking terror' Mount Momotombo, and promptly exploited by the Panama team when it sent each Congressman a letter featuring a Nicaraguan postage stamp prophetically depicting this very mountain furiously fuming away.[30] Coupled with horrific reports of what proved to be the twentieth century's deadliest eruption (that of Mount Pelée on the Caribbean island of Martinique),[31] and wilfully overlooking the fact that the French project had been halted by a major earthquake twenty years earlier,[32] the choice had become clear.

In the wake of the United States' recent military victories as well as its unforeseen audacity to challenge British territorial claims in South America,* US confidence as an international player had never been higher.[33] As a proud social Darwinist and

---

* Specifically regarding the boundary between Venezuela and British Guiana (now Guyana), which remains contested to this day.

disciple of James Monroe's school of geopolitical thought,* Roosevelt believed his country was the rightful leader of the Western Hemisphere, and recognised that by building the canal that had proved beyond even French capabilities, and bequeathing the world (though primarily his compatriots) a convenient inter-oceanic waterway, his country would be immediately thrust towards the top of the international pecking order.[34] Still, before the United States could contend with Panama's near-unassailable natural barriers, it needed to overcome a myriad of diplomatic and administrative obstacles.

First, in order to disentangle Panama from 'the Bogotá lot of jackrabbits'[35] (Colombia was understandably hesitant to lease Roosevelt a coast-to-coast strip of land), the United States stirred up a quick, nearly bloodless revolution, dispatching warships to both of Panama's coasts and bribing Colombian soldiers to lay down their arms.[36] Second, to acquire the necessary land corridor as well as the French assets in Panama, the United States conspired with a Compagnie engineer who, having helped conceive and execute both the Nicaraguan stamp gambit and the US-led insurgency, had already substantiated his commitment to developing an inter-oceanic water connection at all costs.[37] Through the clearly imbalanced Hay–Bunau-Varilla Treaty of 1903, the United States gained the rights to and authority over a roughly 10-mile-wide Panama Canal Zone cutting across the narrow isthmus, as 'if it were the sovereign', 'in perpetuity'. If this immediate infringement of Panama's new national sovereignty wasn't enough, the treaty also granted the United States 'all the rights, power and authority' to use, occupy and control any lands and waters outside this Canal Zone as and when necessary for the

---

* According to the Monroe Doctrine, from 1823 a pillar of the country's foreign policy, European interference in the affairs of a Western Hemisphere nation (i.e. in the Americas) represented a potentially hostile act against the United States.

'construction, maintenance, operation, sanitation and protection' of the canal, and to acquire assets and maintain public order in Panama City and Colón, both cities officially outside the Canal Zone's boundaries, if necessary.[38] Momentarily discounting the not so small matter of physical geography, as far as Roosevelt was concerned, this once obscure corner of Colombia was now primed for 'one of the future highways of civilization'.[39]

The fact that the United States managed to build this quite literally ground-breaking structure over the course of little more than a decade is all the more impressive when one considers that its first year in charge was something of a write-off. Appearing to have learned nothing from the French experience, the original chief engineer John Findley Wallace proceeded by developing the sea-level canal even de Lesseps had accepted was doomed to fail. With thousands of workers dying from disease and most of the rest choosing to return home, progress on the canal was lamentably slow.[40] Just a year into the job, Wallace felt compelled to resign, frustrated by bureaucratic procedures and unwilling to tolerate either the risk of epidemics or the dilapidated infrastructure left behind by the French.[41] Already, US control of the project looked to be an embarrassment, with the manipulation of Panama's natural environment proving too formidable for even this most ambitious and assured of nations.

However, the tide quickly turned with the appointment of Wallace's successor, John Frank Stevens, the civil engineer responsible for extending the Great Northern Railroad across some of the United States' most challenging Rocky Mountain passes. Recognising that a happy workforce is a productive workforce, Stevens, aided by the US Army physician William C. Gorgas and his direct experience of controlling deadly mosquitoes in Cuba,* implemented a remarkably comprehensive sanitation

---

* It was in Cuba that the epidemiologist Carlos Finlay posited in 1881 that yellow fever was transmitted by mosquitoes, although his theory only

programme to mitigate the threat of disease and improve workers' living conditions.[42] The expense involved – well over two-thirds of a billion dollars in today's money[43] – was nevertheless invaluable to the project's eventual success. New procedures such as draining standing water, spraying insecticides and oil, installing screens in windows and building piped-water systems to eliminate the need to collect rainwater in barrels (which attract insects), enabled yellow fever to be eradicated from the zone in around a year, and malaria cases to drop sharply.[44]

In what was once dismissed as inhospitable jungle, new towns duly began to emerge, offering all the infrastructure one would expect of a liveable community, from schools and hospitals to post offices and restaurants.[45] The Americans, once as divided as the French over the optimal design for the canal, also started to implement a more coherent plan. The final days of Wallace's tenure had prompted some necessary introspection, and although most of the Isthmian Canal Commission responsible for overseeing the canal's construction remained committed to building a sea-level canal (albeit in a modified form), a vocal minority, including Stevens, advocated a lock canal instead. In the end, the decision was made in Washington once again. Reports in favour of either argument were considered first by Roosevelt and his hand-picked successor, the secretary of war William Howard Taft, and later by Congress. Agreeing that a lock canal would be cheaper and quicker to build, less expensive to operate and maintain, prove less vulnerable to river flooding and offer quicker and safer passage for ships, as well as being easier to enlarge if desired in future, Congress decided in June 1906 in favour of Stevens's proposal, helping to save the canal project and pre-empting its eventual expansion, to boot.[46]

---

became widely accepted around the start of the twentieth century. The role of mosquitoes in malaria transmission was identified around the same time (separately) by Alphonse Laveran, Ronald Ross and Giovanni Grassi.

Indeed, even in spite of Stevens's resignation just two years into the job, overwhelmed by both political wrangling and the hardships posed by Panama's harsh natural environment, the Americans now had a clear vision for what would be far more than just a conventional canal.[47] In line with Stevens's preference – which happened to be nearly identical to the plan first proposed by de Lépinay but rejected by de Lesseps – the Chagres was dammed to create the world's largest artificial lake, Lake Gatún.* Thanks to a series of locks at Gatún (on the Caribbean side) and Pedro Miguel and Miraflores (on the Pacific side), ships would be able to rise 26 metres (85 feet) from the Caribbean or the Pacific as if climbing an 'aquatic staircase'.[48] From here, they would traverse both the lake and the Americas' continental divide along a water bridge, before descending via locks on the other side. Building such a canal would require additional engineering endeavours. The Panama Railroad[†] was renovated (and in many parts, rerouted) to accommodate more substantial freight, enabling as much as 200 trainloads of debris a day to be carried far away from excavation sites where it might otherwise slip back in.[49] Instead, much of it was used to complete the new lake's dam: thus a potentially hazardous, undesired natural material was turned into an essential resource for a human construction.[50]

To ensure that nature's fiercest currents and winds as well as the material they carry couldn't take vengeance on the canal, enormous breakwaters were constructed at both of its entrances. Dynamite was used to blow through large stretches of Panama's landscape, while a new generation of dredgers, featuring rotating cutters and suction pumps, enabled the American team to break

---

* Depending on whether one is primarily interested in volume or surface area, this title is now held by Lake Kariba (Zambia and Zimbabwe) or Lake Volta (Ghana), respectively.
† Although the creation of Lake Gatún flooded part of the original route, the line still exists today as the Panama Canal Railway.

down and shift a wide variety of rocks and soils.[51] A project that had seemed impossible on at least two occasions was now coming into fruition, a technological imposition on a recalcitrant landscape that had shown barely a mark of human activity merely years earlier. During the first overseas trip by a sitting US president, a photo of Roosevelt controlling a US-manufactured steam shovel – an excavating machine capable of moving earth three times quicker than any of the equipment available to the French – came to symbolise the onset of this new era, characterised by the burgeoning technological prowess and geostrategic influence of a country increasingly determined to mould natural landscapes and international affairs alike.[52]

Like the French before them, the American experience of constructing the canal was not always harmonious, however. The new man in charge, the tireless army engineer George Washington Goethals, brought a leadership style both military-precise and autocratic, and readily fined, jailed and even deported strikers. No less jarring was the intentional creation of an all-encompassing and increasingly stringent system of discrimination and segregation consistent with much of wider US society at this time, whereby workers were divided into 'gold' and 'silver' payroll categories. Inevitably, white American citizens were considered gold-roll personnel and accordingly enjoyed not only superior payment but also better living quarters and food, access to social infrastructure such as clubhouses, and generous holiday time. By contrast, the majority-Black silver-roll workers, most of whom had arrived from Antillean islands such as Barbados and Jamaica,* typically lived in squalid housing, were barred from the eateries and leisure opportunities available to their gold-roll counterparts, enjoyed no paid holiday or sick leave, and were placed under strict curfews to dissuade socialisation and unionisation. What this blatantly racist

---

* Many of the descendants of these workers still live in Panama, which is why this country is home to the largest Black population in Central America.

policy failed to account for was the existence of a third significant group of workers from southern European countries such as
Spain, who, despite regarding themselves as equal to white Americans, were placed on the silver roll as 'semi-white' Europeans.
Incensed by their liminal existence and abusive treatment in a
territory their ancestors had once controlled, Spaniards became
the most willing and anarchical labour activists, often pitting
themselves against both the white authorities and the alternative
'semi-white' workers from Italy periodically employed to replace
them.[53] In this strained climate, one of the few aspects uniting
the workers was a watchful eye for any appearance of Goethals'
unusually vivid motorcar, nicknamed the 'Yellow Peril', which he
used to dash throughout the zone on its railway tracks to check
on the progress of each and every section.[54]

Under Goethals' demanding oversight, the majority of the
canal was constructed rapidly and in accordance with his exact-
ing standards.[55] However, one section continued to vex the project
and illuminated the colonel's general disregard for human life.
To cut through the rockface at Culebra, the point at which the
canal would cross the continental divide, deep holes would first
need to be drilled, then filled with dynamite. Once ignited, a
small chamber would remain, which could then be expanded
with further blasts.[56] Notwithstanding the obvious, direct risks
to human life associated with this unpredictable technology,
each detonation was liable to provoke landslides and mudslides,
imperilling every worker on site, as well as the railway tracks in
the vicinity. At Cucaracha, slides became such a frequent occurrence that almost as quickly as a gap was created in the hillside,
it was filled in again by new material, necessitating further hot,
exasperating and dangerous work.[57] On each day that the mostly
Black workforce tried to move mountains, they faced a real and
constant risk of death that was seemingly immaterial to the
project's leadership. After all, the labourers had come from as
far afield as China and Greece to work on the canal; should

new workers be needed in future, there would surely be a ready supply available.[58]

Finally, nine years and (officially, at least) 5,609 deaths later, the US-led team succeeded in carving through the continental divide, leaving behind an artificial valley through which the canal now runs.[59] The canal's inauguration on 15 August 1914 with the passage of the SS *Ancon*, a steamship renamed in honour of a site near the canal's Pacific terminus, marked what appeared to be the conclusion of a construction project that had become the most expensive in US history – despite taking place outside the then forty-eight contiguous states.[60]

Yet for such a momentous achievement, the Panama Canal barely caused a splash in most of the world when it opened. With German troops advancing across Belgium, much of the media was understandably distracted by a very different, globally relevant event. The canal would fare no better over the next six years, being forced to close for a good 250 days due to a series of landslides (mostly at the Culebra Cut) and labour action, all while international shipping took a hit from the impacts of the war, which taken together meant that the canal could not open to commercial traffic until as late as July 1920.[61] Even San Francisco's world's fair in 1915, described as the Panama–Pacific International Exposition, opted to focus as much on celebrating the city's rapid recovery from a catastrophic earthquake under a decade earlier as on showcasing an infrastructure project that conveniently shaved nearly 15,000 km (8,000 nautical miles) off the traditional sea path through southern Chile's Strait of Magellan or around its Cape Horn.[62] From this immediate low, confidence in the new route could surely only increase. A new era of international connectivity was on the horizon. The rest of the world just needed a moment to recalibrate itself.

For most of the twentieth century, the United States' construction and subsequent administration of the Panama Canal represented a central aspect of its burgeoning reputation as the planet's pre-eminent state. In proving its capacity to surmount Panama's natural obstacles and develop a maritime passageway to be used, theoretically at least, by a ship from any nation with ocean access, the United States could plausibly claim to have offered international society advantages that would have been unthinkable mere decades before. From the accelerated transit of consumer goods to foreign markets, to safer journeys for navigators crossing between two oceans, the world had never been so connected. Although the canal's military significance declined rapidly after the Second World War for various reasons,* its earlier role in allowing the United States to become one of the world's major naval powers, capable of quickly redeploying ships between the Atlantic and the Pacific, should not be overlooked.[63] Powered by the flair and gumption of the Technological Revolution, the canal helped the United States to readily expand its economic and political power beyond its own borders, and position itself at the core of our modern, globalised world.[64] However, the story of the Panama Canal pertains not only to the macro scale of connection, of new trading opportunities and the start of a new era in world history. When we look closer, a sad irony appears: for several decades, the country that stood to gain the least from the Panama Canal and the international connectivity it facilitated was Panama itself.

For Panama, US interest in the canal project has often proved bittersweet. The country's very independence owed much to the United States' growing tendency to involve itself in foreign affairs, yet that same interventionism spurred the US to treat Panama more

---

* Among them, growing concern that a hostile nation would deem the waterway an obvious target (keen to ensure its safety, in 1945 US president Harry S. Truman even suggested that the canal be handed over to the newly established United Nations instead), the development of new aircraft carriers that were too large to fit through its narrow locks, and evolutions in military technology that saw face-to-face naval encounters become almost obsolete.

like a colony and a canvas for its legislative and industrial power than a country in its own right. On top of affording the United States near-sovereign authority within the 10-mile Canal Zone, the Hay–Bunau-Varilla Treaty allowed the seizing of additional Panamanian territory, should it be deemed necessary to the canal's effective functioning. Panama was also forbidden from construct-ing its own trans-isthmian communication systems. Meanwhile Panama's constitution, drafted with foreign interests patently in mind, gave the United States the right to intervene in any part of Panama where it judged the canal under threat. Together, these documents enabled the United States to expand Panama's canal infrastructure and its political control thereof simultaneously, all under the guise of protecting the canal from harm.

It is easy to understand many Panamanians' resentment towards what was only nominally their canal. While the new route plainly benefited the United States economically, boosting its trade so consequentially that it wasn't motivated to charge higher tolls for foreign vessels,[65] for the Panamanian govern-ment this not-for-profit development deprived their country of much-needed revenue.[66] To make matters worse for the Panama-nian government's coffers, and reflecting how connections within Panama seemed to operate in only one direction, the sanitation systems constructed by the Canal Zone administration in Panama City and Colón were charged to it with interest, while Panama-nian businesses were prohibited from selling goods in the Zone or to vessels passing through.[67] Alongside an executive order by President Taft that forced around 40,000 working-class people to abandon their homes soon after the canal's completion, and con-tinued discrimination against Panamanian labour in the region, directives such as these had the effect of turning the Canal Zone into an exclusive exclave, a tropical idyll for US citizens which Panamanians were discouraged from re-entering.[68] Meanwhile, for Panama's Indigenous communities, the creation of Lake Gatún and, in the 1930s, a second reservoir at Alajuela, outside the Canal Zone, involved flooding tracts of their lands.[69] For many people living in Panama, the canal signified not connection with the wider world, but detachment domestically.

Other nations similarly tended to view the Panama Canal as a US structure on Panamanian territory: Japan, for instance, whose 1945 plan to destroy the locks using submarine aircraft carriers was aborted at the last moment, mere weeks before the atomic bombings of Hiroshima and Nagasaki brought the Second World War to a close.[70] Panama did manage to achieve some minor amendments to the Hay–Bunau-Varilla Treaty – most significantly in 1936, when the United States relinquished its protectorate role and its right to intervene in Panama's internal affairs, and in 1955, when President Dwight D. Eisenhower granted Panama the right to tax its citizens working in the Canal Zone. But in general, Panama was unable to escape being a client state to the world's most powerful nation, a non-contiguous extension of the United States whose own territory was split in two.

No place symbolised Panamanians' marginalisation in their own country more than the Canal Zone – an issue magnified in 1959 when this privileged strip of land was fenced off, giving credence to its deplorable motto 'Panama divided, the world united'.[71] Fittingly, the impetus behind the decision to physically partition a country whose original *raison d'être* had been to guarantee connection was a piece of fabric with the capacity to both divide and unite. On the occasion of Panama's Independence Day (3 November), celebrations turned to demonstrations and then violence when Panamanian protesters attempted to raise their flag in a territory they considered theirs.[72] In a pattern replicated over the next two decades, the Americans were split. Whereas some, including Eisenhower, were willing to make certain symbolic concessions to Panama, including recognition of its titular (as opposed to actual) sovereignty over the canal, and called for the flying of both nations' flags in the Zone, most of his compatriots both in Washington DC and in the Zone opposed the very notion of ceding any sort of control to this small Central American country.[73] Ultimately, the erection of Panama's version of the Berlin Wall only materialised the unhappy populace's sense of detachment, by presenting them with a tangible reminder of who was really in charge.[74]

Still, reminders can represent a double-edged sword. Though the 'Fence of Shame' perturbed many Panamanians, it and the flag dispute behind it simultaneously handed them overt symbols of division and disparity against which they could direct their ire. In particular, the Canal Zone's choice of flag remained a significant source of antagonism, involving a tempestuous melange of opinions among Panamanians, Americans and Zonians* alike. As Zonian students protested their governor's sudden refusal to allow the US flag to fly outside their schools – a decree at odds with the recently assassinated President John F. Kennedy, who had believed that both flags should stand together as a symbolic statement of unity[75] – Panamanian students saw an opportunity to finally claim sovereignty over the Canal Zone. Marching into the exclave at the end of classes, they approached Balboa High School, where the US flag had been repeatedly lowered and hoisted over the preceding two days, and demanded that their flag be raised in its place. Despite some Zonian sympathy for their cause – the district's head of police, for instance, offered to allow a small delegation of the Panamanian students to display their flag at the foot of the school's flagpole and sing their national anthem – this unwelcome assertion of Panamanian sovereignty duly provoked a scuffle

---

* Though far from monolithic, Zonian identity generally fuses US and Panamanian national identity, most individuals having descended from the US citizens who constructed and subsequently maintained the canal during the Canal Zone's existence. Arguably no Zonian was better known than John McCain, a US senator for Arizona for over thirty years who ran, unsuccessfully, as the Republican presidential nominee against Barack Obama in 2008.

between students, adult civilians and police, during which the Panamanian flag was somehow torn. As news of the apparent flag desecration quickly spread throughout the country, Panamanian demonstrators began to assault American cars and buildings with rocks and incendiaries, and tore down sections of the detested fence. The Canal Zone's security forces hurled tear-gas grenades and opened fire on the protesters in response, in many cases indiscriminately. Today commemorated in Panama as *Día de los Mártires* (Martyrs' Day), this tumultuous day, 9 January 1964, would see the controversial deaths of twenty-one Panamanians* and four US soldiers.[76] The image of three men climbing a lamp-post to raise their nation's flag, while a burning car sends thick smoke into the air behind them, remains one of *Life* magazine's most visceral covers.

With US legitimacy at an all-time low in a region that, owing to its internationally connective power, was supposed to be fundamental to US foreign policy, the Panamanian president Roberto F. Chiari made the unprecedented move of briefly breaking diplomatic relations with the country that had brought it theoretical independence six decades before.[77] However, new negotiations yielded little until the ascent of a political outsider to the US presidency, on a platform that pledged to recalibrate his country's moral compass and achieve stability in a world whose growing fragmentation demanded new sources of unity. Before rejuvenating a stalled peace process in the Middle East with the Camp David Accords of 1978, Jimmy Carter made Panama a key focus of his tenure, signing two treaties in 1977 with Omar Torrijos, the de facto leader of a country that had descended into military dictatorship.[78] These agreements marked a sea change

---

* Certain details surrounding this distressing event are hazy, but the victims appear to have included a six-month-old girl caught in tear gas in Colón, and an eleven-year-old girl hit by a stray bullet while standing on her family's balcony in Panama City.

in US–Panamanian relations. In addition to terminating all prior treaties between the two countries regarding the canal, the Canal Zone, the basis of so much discord and division, would be abolished from 1 October 1979. To ensure a smooth transition process to full Panamanian sovereignty, a joint US–Panamanian agency would replace the existing US government-directed corporation in charge of the Canal Zone's governance until 31 December 1999, when Panama would acquire full responsibility for operating the canal and maintaining military forces and installations within the entirety of Panamanian territory. Further, Panama would guarantee the permanent neutrality of the canal even in times of war, meaning that rather than the canal constituting a central node of a necessarily US-oriented world, from this point on it shouldn't automatically privilege any one country over another.

At long last, Panama was poised to liberate itself from direct US oversight and allow the canal to work truly to its own advantage. There was, however, still time for one more US intervention. Though Torrijos's relationship with the United States wasn't always cordial, his sudden demise in a plane crash in 1981 allowed his even more controversial ally Manuel Noriega to rise to the peak of Panamanian politics.[79] As a keen conduit of goods and information, Noriega essentially personified the Panama Canal, and in the country's fight against communism, the United States often valued his willingness to share aid, arms and intelligence with its associates in Nicaragua and El Salvador.[80] Unfortunately for the superpower of the capitalist world, Noriega's support was never motivated by ideology: as a true opportunist, he was just as willing to assist hostile nations such as Cuba as he was to build relationships with US leaders, all while cultivating intimate links to Colombian drug cartels.[81] As his leadership took an increasingly repressive bent, it became harder and harder for the United States to countenance any further partnership with a man whose criminal activity risked undermining the canal's reputation, and whose forces were now harassing US military personnel and civilians.[82]

And so, when insistent words and the withdrawal of economic and military aid failed to pressure Noriega to stand down, in late 1989 the United States opted to initiate its last assertive hurrah in Panama. Contending that democracy, US citizens' lives in Panama and the very integrity of the Torrijos–Carter Treaties were all threatened by a military dictator who had turned the country into a loathsome hub of drug trafficking, President George H. W. Bush's forces succeeded in chasing down and ousting a former ally in a matter of weeks, via a highly unorthodox method.[83] Having learned that this notorious drug smuggler with a penchant for prostitutes was hiding out in the unlikely confines of the Holy See's diplomatic offices in Panama City, the Americans' successful strategy involved blaring out a playlist of rock anthems with a common theme: 'Manuel, your days in charge are numbered'.* Though many international observers were outraged by what they viewed as a flagrant violation of Panama's sovereignty and international law – the invasion, that is, not the refrains of 'No More Mister Nice Guy' by Alice Cooper or 'Wanted Dead or Alive' by Bon Jovi – and the United Nations General Assembly condemned the invasion by a vote of seventy-five to twenty, few in Panama seemed to care.[84] Finally, Americans and Panamanians appeared to be on the same page, assured that with the strongman out of the picture, the connective infrastructure the United States had built and managed according to its own interests could now work to Panama's benefit as well.

For over a century, the Panama Canal has stood as one of the world's greatest symbols of international connectivity, a transit route that provides greater convenience to shippers than nature could offer alone. A territory with negligible natural resources but an unusually linear configuration between two oceans, Panama

---

* A further advantage of this passive-aggressive scheme was that the media wouldn't be able to overhear the details of the Americans' negotiations with the ecclesiastical office.

owes its independence (both de jure and de facto) in large part to its potential commercial and geopolitical advantages: its physical geography has been exploited to guarantee its very place in the world. Panamanians constantly encounter subtle reminders of their country's internationally oriented, connective significance: the canal joins two places named after European explorers (the Italian explorer Cristóbal Colón, better known in English as Christopher Columbus, and the Spanish conquistador Vasco Núñez de Balboa), while its oldest bridge, Puente de las Américas, belatedly reattaches the two halves of the Americas. More pragmatically, through capitalising on its newfound freedom to run the canal as a profit-making enterprise, Panama today enjoys a significant economic boon over its regional competitors, and additionally, owing to a period of political stability matched only by Costa Rica, it has quickly become one of the fastest-developing countries not only in Central America, but globally.[85] Panama might not have felt the benefits of the canal immediately, but now this infrastructure is part and parcel of its present and future.

At the same time, billions of people across the world continue to rely on the canal, generally without even realising it. By easing trade between the planet's two biggest oceans, the waterway has long supported the rapid economic expansion not only of the United States, but of Japan, South Korea and Taiwan across the Pacific as well.[86] On a daily basis, commodities including motor vehicles, petroleum, liquefied natural gas, coal and grains all pass through a canal that has welcomed more than a million vessels since opening little over a century ago.[87] Employment in the shipping industry depends in no small part on a waterway that each year conveys around 40 per cent of US container traffic alone.[88] Today it is impossible to imagine the planet without this critical, human-made conduit, which acts as both a barometer of the world economy and a facilitator of economic growth.

The canal remains a testament to ingenuity and perseverance, its multiple engineering features epitomising a very human determination to control and modify the planet so that it works ever so slightly more in our favour. Its locks are carefully arranged so that water from Gatún flows by gravity to these essential components of the canal system, while the same lake's dam helps supply the country with its biggest source of electricity, hydropower. The variable Miraflores locks are so designed because the Pacific Ocean sees larger tidal ranges than the Caribbean Sea[89] – an issue de Lesseps' team would have eventually encountered had he been able to pursue his sea-level canal plan. Because of human actions, the Chagres, the fast-flowing river that had once posed a formidable obstacle to the canal's construction, now bears the unusual distinction of draining into two oceans. To maximise the canal's navigability, specially trained canal pilots assume responsibility for manoeuvring each boat expertly from one end of the waterway to the other; at the locks, the largest vessels are additionally assisted by teams of electric locomotives using strong cables and robust winches. Meanwhile, in other places, human dominion over the natural landscape is legislative as well as technical: in creating Gatún, which resembles an enormous puddle on a map of Panama, the Americans turned a high hilltop at Barro Colorado into a remote island, which they subsequently designated as a nature reserve. Now under the administration of the Smithsonian, Barro Colorado Island claims to be the most intensively studied portion of tropical forest in the world.

What's more, in encapsulating how our relationship with the planet is dynamic rather than static, the story of the Panama Canal and our modification of the environment necessarily remains unfinished. Due to our demanding expectations as consumers and the general growth of the shipping industry, first the United States and now Panama have had to embrace the ceaseless challenge of implementing new ways of guaranteeing ever-increasing numbers of container ships (whose dimensions have long been determined by the canal) quick and convenient passage from one side to the other.[90] The installation of fluorescent lighting in the 1960s allowed the canal to offer 24/7 access, while in the 1990s, the widening of the old menace the Culebra Cut, supplemented by updates to the canal's lock technology, helped ease the transit of Panamax vessels – the maximum size capable of passing through.[91] Even more momentously, having gained full control of the canal at the start of 2000 and therefore no longer needing to answer to US priorities, an autonomous government agency, the Autoridad del Canal de Panamá (ACP; Panama Canal Authority), oversaw the country's biggest infrastructure project since the canal was built in the first place. After overcoming some significant cost overruns, delays and uncertainties as to its eventual completion,[92] two new lock complexes were constructed parallel to the existing locks to add a new lane of traffic, and the Caribbean and Pacific entrances as well as Gatún were widened and deepened.[93] As a consequence, not only is the canal's capacity now far greater, allowing it to accommodate post-Panamax vessels capable of carrying nearly three times as much cargo as their Panamax analogues, but the country's usable water reserves – aided by the new locks' water recovery system – have been boosted, too.[94]

Thanks to this project, the Panama Canal has only become more essential to international trade.[95] At present, the average vessel can cross the Panama Canal in around ten hours compared to almost a day prior to the expansion (even if the two-hour, forty-one-minute transit recorded by the US Navy hydrofoil *Pegasus* in 1979 proves that it was already possible to travel far faster, as long as you have the necessary permission), and owing to the new possibility of leading two ships through the canal simultaneously, waiting times at either entrance should theoretically be reduced.[96] Keen exporters such as China are now able to send even larger ships via this route, so that masses of its goods quickly reach distant markets in North America and Europe.[97] Although the canal remains too narrow for some super-sized cargo vessels, including the Triple E-Class container ships used elsewhere in the world by Maersk Line, this and other major shipping conglomerates also tend to favour the route because of the time, money and carbon dioxide emissions it saves port to port.[98]

Nonetheless, soon enough, humans must pay a price for modifying and shaping natural landscapes to meet our desires. With both 2016 and 2023 bringing drought to Panama, important questions have been raised about public water consumption in a country whose capital's glittering skyline evinces its rapid recent development, and whose canal, despite its new freshwater-saving features, drains well over seventy Olympic-size swimming pools per vessel.[99] By the end of 2023, the canal's water levels fell so low that instead of the usual thirty-six to thirty-eight vessels per day, just twenty-two were allowed to pass, many at low capacity lest their weight cause them to scrape the bottom, forcing them to send their surplus across the isthmus using the rail route that had once helped the canal materialise in the first place.[100] Facing intolerable delays and the uncertainty these bring to an industry reliant on predictability, some shippers opted to make risky journeys via routes the Panama Canal was supposed to have rendered obsolete (for instance, the narrow and gusty Strait of Magellan), as well as its long-standing rival the Suez Canal in Egypt.[101] Others decided that such perils weren't worth the risk, preferring instead to sit tight for long, monotonous days at the canal's entrance.[102] However, as is almost always the case, money speaks – especially fossil fuel money. In paying nearly $4 million simply to jump the line, not including the mandatory transit fee all users must pay to travel through the canal, the Japanese petroleum and metals conglomerate ENEOS[103] encapsulated many shippers' modern-day dependence on a canal that charged the US travel writer and adventurer Richard Halliburton just thirty-six cents* to swim its length back in 1928.

On other occasions, the issue has been not a lack of water, but too much. Whereas the canal's rare previous closures tended to

---

* Adjusting for inflation, about $6.50 today.

be compelled by intermittent landslides – the obvious exception being the closure during the 1989 invasion to depose Noriega – December 2010 saw the canal shut for seventeen hours, amid heavy rains that raised the water of Gatún and Alajuela to alarming levels.[104] Given also the continual need to dredge channels and monitor the stability of the Culebra Cut's slopes in the wet season, one can say that waterways such as the Panama Canal offer convenience to a point. It's one thing to shape the earth, quite another to control it.

Particularly in this part of the world, a defective canal spells economic disaster. While Panama receives less toll revenue when the canal is underutilised, ports from El Salvador to Chile also experience consequential reductions in maritime trade, most notably with the US East Coast.[105] For this reason, alternative initiatives to connect the Atlantic with the Pacific have been considered in recent years, including a revival of the old Nicaragua Canal project and a new rail link across Colombia, both with Chinese investment.[106] For the meantime, though, the two coasts of the Americas remain quite surprisingly disconnected. In Panama itself, various solutions to the growing water crisis have also been proposed, including the construction of a new reservoir on the Indio River to supplement the canal's freshwater reserves.[107]

However, major infrastructure projects can be a highly sensitive matter. Indigenous communities including the Kuna and Emberá have already experienced the trauma of infrastructure-related displacement, and it is understandable that many are resistant to new programmes that necessitate the total transformation of their lands.[108] For some *campesinos* (peasant farmers), too, the canal is closely associated with suppression and dislocation: many were fined and jailed for engaging in traditional swidden agriculture (also known as slash-and-burn agriculture or shifting cultivation) during the Noriega era, on the basis that it was causing the canal to silt up.[109] Forced onto smaller and smaller plots over time, all while cattle ranchers and miners have frequently received government support in spite of their own environmental impacts, some continue to resent a waterway that enjoys greater protections than human lives, livelihoods and landscapes.[110] After all, in the process of modifying Panama into a nation that works 'for the benefit of the world',* cherished lands and historic settlements have been wiped away, replaced by a transport infrastructure that still prioritises international shippers over many of Panama's residents, to say nothing of a former Canal Zone that excluded Panamanians and physically disconnected west from east.[111] Even the elimination of the Zone has failed to truly integrate a country that effectively comprises a wealthy, urbanised strip of land following the canal through the centre, surrounded by two poor, underdeveloped rural segments that remain largely detached from economic activity beyond their boundaries.[112] In sum, the canal has never been beloved by or profitable to every Panamanian, and so just as nature may fight back, there is always the risk that people will do so as well.

Perhaps inevitably, considering its global relevance, Panama's challenges are not merely internal. Most pertinently, as long as Panama holds vital keys to the global maritime trade, it is compelled to negotiate the geopolitical pressures exerted by rival nations, rendering it a key arena of modern diplomacy. Unsurprisingly, as both the two biggest users of the canal and the contemporary world's two foremost (and rival) economic powers, the United States and China pose a particularly significant predicament, and in recent years different Panamanian leaders have disagreed intensely on Panama's relationship with each nation. Before 2019, President Juan Carlos Varela forged close ties to China, cutting off his country's diplomatic ties with Taiwan and signing Latin America's first 'Belt and Road' agreement, as well as hosting Chinese president Xi Jinping for an official visit in December 2018. Panama looked to be on the precipice

---

* *Pro Mundi Beneficio*, Panama's unusually humble and candid motto.

of becoming a transit wonderland, with a new bridge over the canal, a high-speed rail line from Panama City to David in the country's far west, a modern cruise ship terminal in Panama City, and an advanced container port in Colón all set to be built with Chinese money.[113] If these projects didn't alarm the United States enough, then the plan to establish a giant new Chinese embassy at the canal's Pacific mouth certainly did, causing US officials to pressure Varela to withdraw his offer of a plot of land in this unmatched location, before a massive red flag with five gold stars could greet each and every user of the canal the US had built.[114]

Then, all of a sudden in 2019, the winds of change converged on Panama. Under Varela's successor Laurentino Cortizo, multiple Chinese projects were suspended or cancelled, purportedly due to the failure of some Chinese companies (which in reality often have close ties to China's political leadership) to comply with their commitments, but just as importantly because of US diplomatic pressure to prevent China from running Panama's ports and thereby undermining the neutrality Panama had promised to uphold back in 1977.[115] Panama's withdrawal from the Belt and Road Initiative in February 2025, amid accusations from the second Trump administration that the waterway had fallen under Chinese control, has since provided an additional reminder that the canal remains as relevant to the United States today as it was a century ago. The only key difference is that the parties that could potentially challenge US influence have changed.

It's evident here that retaining a degree of authority over the canal is not simply a practical issue for the United States; it is also a symbolic concern, evoking the tight bond that has existed between country and canal for well over a century. For France, caught up in the euphoria of completing the Suez Canal, the Panama project had been financially motivated in the main: its proponents called on citizens of all social classes to invest money and, all going well, become rich. But for the United States, building the canal (with government funds) was, to a far greater extent, a triumphant declaration that the New World was the future, at the same moment as the Old was splitting apart.[116] By completing a project beyond the capabilities of the country that had built the world's tallest structure and constructed the planet's other pre-eminent canal, the United States could now reasonably boast that it was modern society's leading engineer, with the unique ability to effectuate a vision of a more interconnected and convenient world.[117]

In combining steel from Pittsburgh with workers from Barbados, coordinated by a well-travelled and experienced military leadership, the Panama Canal marked the emergence of a muscular and globally oriented United States, a budding imperial power with the desire and ability to tame nature and rival states alike. Until *Apollo 11* in 1969 achieved the remarkable feat of landing humans on the moon, this megaproject arguably represented the United States' greatest and most internationally relevant engineering accomplishment. Comprising not just a channel but also locks, dams and railway tracks, and relying upon novel dredging and blasting techniques, it is easy to see why the Panama Canal was designated as one of the Seven Wonders of the Modern World by the American Society of Civil Engineers in 1994, and of this exclusive group,* it is conceivably the only one with real, practical importance to the majority of people worldwide. The canal is what most people associate with Panama first and foremost. This megaproject more than put the country on the map: it positioned it at the heart of everyday life. Now Panama is fully in charge of its own affairs, the challenge is to stay there, and flourish.

---

* The other six being the Channel Tunnel, the CN Tower, the Empire State Building, the Golden Gate Bridge, the Itaipu Dam and the Netherlands North Sea Protection Works.

4

Reimagination: THE LINE

———————

THE LINE will tackle the challenges facing humanity in urban life today and will shine a light on alternative ways to live.

Mohammed bin Salman[1]

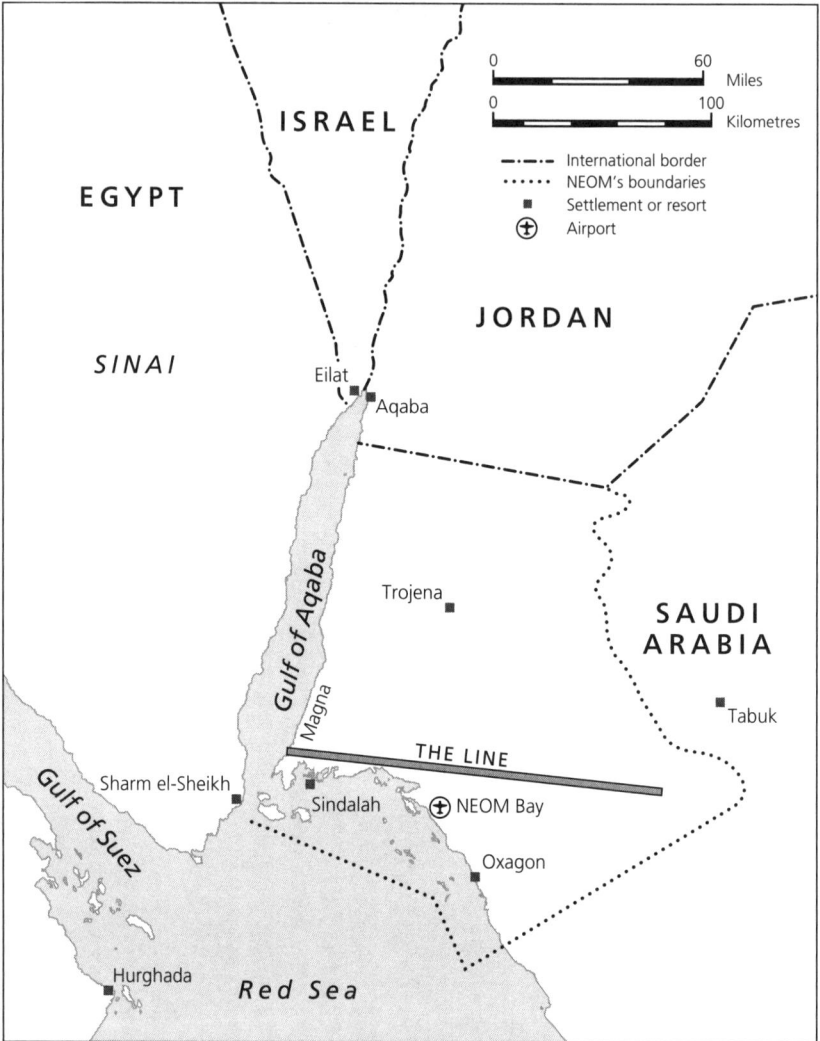

ISRAEL

EGYPT

SINAI

JORDAN

Eilat

Aqaba

Trojena

SAUDI
ARABIA

Gulf of Aqaba

Magna

Tabuk

THE LINE

Sharm el-Sheikh

Gulf of Suez

Sindalah

NEOM Bay

Oxagon

Hurghada

Red Sea

0   60
          Miles
0   100
          Kilometres

— · · — International border
· · · · · · · NEOM's boundaries
■ Settlement or resort
⊕ Airport

These days, everyone seems to have an opinion on Saudi Arabia. Golf enthusiasts grapple with the emergence of a Saudi-financed and Trump-approved tournament to rival and ultimately absorb the classic PGA Tour, while fans of Europe's elite football clubs chastise a growing cohort of players seduced by mind-boggling remuneration in a league ranked lower than its Cypriot and Swedish counterparts. Generally from a distance – for organisations such as Amnesty International are banned here – human rights activists and advocates censure a country where violations of sharia (Islamic law), including same-sex and extramarital relationships and cross-dressing, can result in punishments ranging from flogging to execution, where dissent in the form of peaceful protest, free expression or promoting atheism risks a death sentence, and where political opponents and migrant workers can be tortured in detainment. The Yemeni civil war since 2014, in which a Saudi-led coalition has intervened with direct air strikes on civilian targets such as hospitals and schools, has raised important questions internationally about British and US arms exports to Saudi Arabia, while a growing portfolio of other recent Saudi purchases – including stakes in Microsoft, Starbucks, PayPal, Uber and Electronic Arts – is forcing business executives to define their positions on this increasingly influential country. Regardless of whether the subject is a dinner-table taboo or a water-cooler chat, the Middle East's largest nation has moved to the centre of many a conversation.

Still, it's one thing to be discussed, quite another to shape the narrative. Cognisant of this fact, Saudi Arabia is attempting, more and more, to recalibrate how it is perceived by outsiders.

In an effort to rebut popular perceptions of it being an austere hotbed of Wahhabi* conservatism and gender inequality, in 2018 the country renounced its long-standing ban on female drivers and authorised women to leave the house without a veil.† Cinemas, prohibited since the early 1980s on account of their potential to corrupt viewers' morals and promote 'foreign' liberal values, were reopened concurrently, allowing citizens to be inspired by international creations (censorship notwithstanding) and advance a fledgling domestic film industry. In contrast to its long history of educating male and female students separately and prioritising traditional Islamic subjects, in 2009 Saudi Arabia opened its first co-educational university, King Abdullah University of Science and Technology (KAUST), which specialises in urgent contemporary issues such as environmental protection and bioengineering, and hosts students from more than sixty countries. Further exemplifying its interest in shaping the future, Saudi Arabia is today positioning itself as a global leader in artificial intelligence, a place where robots can gain citizenship and pioneering diagnostic systems can accurately detect patients' ailments.[2] In addition, by supplementing its increasingly educated population with foreign talent and expertise, the country is working hard

---

* A puritanical movement within Sunni Islam that stresses literal interpretation of Islam's principal texts. Since 1744, this sect has played a key role in shaping the politics of the territory today called Saudi Arabia.

† This is not to say that Saudi Arabia has suddenly become a liberal, egalitarian country, however. A good example is its wavering policies on male guardianship, a system that has historically obligated women to rely on a male guardian to travel abroad, make pilgrimage to Mecca or register the birth or death of a relative. These restrictions were lifted in 2019, and yet three years later, ironically on International Women's Day, men's unique powers in marriage, in attaining a divorce and in designating child custody were all codified in law. Moreover, even though women are no longer compelled to wear a loose-fitting *abaya* overgarment, they are still forbidden from wearing 'immodest' outfits that show or hug their skin.

to overcome any image of being a parochial, primitive desert nation whose global relevance is limited to exporting the modern world's favourite energy source, demonstrating instead that it can welcome newcomers committed to helping it power and captivate the planet like never before.

To this end, Saudi Arabia's most astounding departure from the past is its sudden commitment to ending its dependence on the finite and economically volatile fossil fuel supplies that made it rich in the first place.* Through investing inconceivable amounts of oil revenue in other industries, most notably leisure and tourism, the country is now striving to offer an unmatched life-style for visitors and expats as part of its 'Saudi Vision 2030', an ambitious programme committed to evolving the country into a diversified and fundamentally turbocharged economic and social hub. Welcoming glampers and other luxury travellers seeking to explore the nearby UNESCO World Heritage Site and archaeolog-ical dreamland of Hegra (Al-Hijr), Saudi Arabia's answer to Petra in Jordan, new boutique resorts as well as a flying museum are transforming and reimagining the practically untouched ancient oasis city of Al-'Ula. Two new tourism megadevelopments – the Red Sea Project and AMAALA – will soon endeavour to attract the world's rich and famous to the country's west coast, while another, Qiddiya, promises a state-of-the-art Formula 1 circuit, a golf course designed by Jack Nicklaus and the world's longest, tallest and fastest rollercoaster. If any type of building screams 'Look at me!', it's surely a skyscraper, and at a kilometre (over 3,000 feet) high, once completed, the Burj Jeddah (Jeddah Tower) will comfortably overtake Dubai's Burj Khalifa as the world's

---

* This political shift from fossil fuel energy has hastened since September 2019, when drones dispatched by Houthi forces in Yemen were used to attack Saudi Aramco's oil processing facilities, temporarily undermining the country's oil production and briefly forcing it to *im*port energy, while oil prices elsewhere spiked by as much as 20 per cent.

tallest and most commanding structure. In Riyadh, the old international airport is being converted into the world's largest urban park, housing amenities as incongruous as a skydiving centre, a virtual reality court, a butterfly sanctuary and an Islamic-style garden, along with six museums and a mixture of residential, office and retail space, while plans are underway to construct a cube-shaped skyscraper called the Mukaab, where virtual reality technology will allow visitors to travel to Mars or a magical, fictional world. Even Islam's holiest city and pilgrimage capital hasn't been exempted from Saudi Arabia's extraordinary rebranding campaign: today it boasts the world's fourth tallest building (the Makkah Clock Royal Tower, which contains a five-star Fairmont hotel and five-storey shopping mall) and what will be, once completed, the world's biggest hotel (Abraj Kudai), alongside manifold luxury residences, commercial spaces and restaurants. Overlaying the traditional with the modern like the *kiswa** covers the Kaaba, this monumental and modish Mecca may be accessible only to Muslims, but it is catching even the most secular of eyes.

Having smashed its 2019 target of welcoming 100 million tourists by 2030, in 2023 the Saudi government raised its goal to 150 million, showcasing its growing confidence and ability to entice and excite.[3] Suddenly, Dubai has a realistic challenger to the unofficial title of the world's most fanciful society. And yet these extraordinary modifications to the country's landscape – some new resorts here and there, a big park, a few skyscrapers – are small fry by Saudi standards. For in its efforts to position itself at the centre of international attention, and by extension, at both the planet's figurative and literal heart, nothing is more intriguing than Saudi Arabia's interest in revolutionising our most significant nexus: the city.

---

* The distinctive black silk cloth draped over the Kaaba in the Masjid al-Haram (Great Mosque) on the ninth day of Dhu al-Hijjah, the last day of the Islamic calendar.

Cities have historically been connective hubs of civilisation, bringing together people and resources from far away, yet perpetually prone to inefficiency, excess and environmental destruction. Recognising this reality, today Saudi Arabia insists that urban areas should be rethought and redesigned almost entirely. In following a blueprint that advocates thorough integration between the human and natural worlds, and incomparable connectivity among urban functions and spaces (for instance, residential, commercial, educational and economic), Saudi Arabia's most radical megadevelopment promises to alter the very essence of our lived existence. Boasting almost unlimited space, resources, imagination and political authority, and conveniently situated at the geographic intersection between Asia, Europe and Africa, no country is better suited to acting as a laboratory for such an innovative and subversive endeavour.[4] For the meantime, the rest of us just need to forget what we think we know about cities. Apparently, for six millennia, we've been doing them all wrong.

Imagine for a moment that you have been commissioned to develop a brand-new city. You likely have some questions for the council to ensure you best realise their vision:

> What should stand at the heart of our new city, a site that will best embody our society's priorities? A civic building, perhaps, or maybe a temple? How about a commanding monument, a public square, or even a beautiful park?
> *We're going to stop you right there. You're going about this the wrong way. Our city shouldn't even have a centre.*

No centre? How odd. I guess that means we won't be arranging the streets like the spokes of a wheel, radiating from a single hub. But a grid system would still work. After all, you can draw streets as a lattice without needing to determine a single downtown area.

Actually, our city won't have streets – they're not necessary. So you can forget about drawing a grid, too.

Wait, what? A city without streets? So how will people travel, and how will they park their vehicles? And if they're not going to a central place, where *are* they going?
They won't need their own cars: they can walk to all local amenities, and jump on a train to travel further afield.

Well that sounds quite forward-thinking, if a little quixotic. But I'm game. Roads are so tedious to name, after all. I mean, there are only so many plants, professions and political figures worth honouring. What about living spaces: surely part of the city will be dedicated to dream homes, detached mansions with expansive gardens?
No chance – everyone, regardless of social status or income level, will live as well as work in one of two skyscrapers.

So a city should essentially be a pair of buildings, with a long railway line, but no roads and no cars?
Bingo.

Confused yet? With a flair appropriate for the digital age, the developers of this city – whose name, THE LINE, is rather more unassuming than one might anticipate of such a supposedly paradigm-shifting enterprise – have produced a nifty website explaining their vision.[5] At its core is a stance both pessimistic and hopeful, pragmatic and fantastical: that in the face of climate change, the modern world's greatest existential crisis, human ingenuity will ultimately prevail. While much of the world bakes and burns, with towns swallowed up whole by fire or sea, and new conflicts erupt over ever-dwindling fresh water and food supplies, here in Saudi Arabia, people are taking control of a planet that, though broken, is not unfixable.[6] Modern life's greatest problems and biggest annoyances will be non-factors here. Air pollution? No more. Traffic congestion – no, worse, traffic *accidents*? Consigned to the past. Imagine no longer having to worry about your carbon footprint, or wasting your life commuting. Out of mobile data, and can't connect to Wi-Fi? A free citywide 5G network, enabled by satellites and high-speed fibre optics, will allow users to connect with anyone, anytime.[7] Never want to do the housework again? Robot maids and butlers will be at your beck and call.

With a élan style that could make the most ambitious of property developers blush, the city's 'placemakers' boast that THE LINE will be a 'cognitive city', a 'mirrored architectural masterpiece', which will constitute a 'civilizational revolution'. Though it's unclear whether even a modicum of inspiration was taken from the Alaskan town of Whittier's single, multi-purpose building, THE LINE's boosters contend that their city, which will comprise twin skyscrapers 200 metres (660 feet) wide but 170 km (105 miles) long, will 'redefine' the very concept of a city and 'what cities of the future will look like'.[8] With a target population of 9 million by 2045 living in an area about half the size of San Marino, the sprawling, low-density model of urban development beloved by so many North American planners could be history.

Key to this new vision is the idea of building cities vertically – what THE LINE's planners call 'zero gravity'[9] – so that people move up and down as well as across space. To achieve this lofty objective, THE LINE's two skyscrapers will contain different 'modules' – that is, mixed-use communities of up to 80,000 people each, on different storeys, rising to a height of 500 metres (1,640 feet),[10] a fact that will place the skyscrapers among the tallest buildings in the world. By maximising the potential of three-dimensional space, THE LINE will allow residents to access everything one would expect of a city by travelling not necessarily horizontally (via metros positioned on different levels), but vertically (by elevator). Should a resident want to watch a new film, say, all they will need to do is travel either up, down or across to whichever cinema-holding module offers the quickest and most convenient journey from their own module. For everyday amenities, such as schools, shops and parks, it won't even be necessary to leave one's own module. By taking Archimedes' maxim that 'the shortest distance between two points is a straight line' to literal new levels, THE LINE promises to exemplify how profoundly connective a city can be.

Nor is THE LINE intended as an isolated linear development, stretching aimlessly across a corner of the country. Rather, it will constitute the axis of a new high-tech conurbation called NEOM, the self-declared 'land of the future',* whose other 'regions' will include a ginormous eight-sided floating port city called Oxagon, a world-class ski resort named Trojena, a luxury Red Sea island resort known as Sindalah and a string of twelve luxury resorts and residences set within a shoreline nature reserve, called Magna.[11] In addition to offering easy cross-border travel to two emerging market economies with growing tourism sectors in Jordan and Egypt, including the resort city of Sharm el-Sheikh via an eventual bridge, the city's location is said to allow travellers to reach 70 per cent of the world within eight hours from its quartet of airports (one is already open), and 40 per cent in just four.[12] Under the leadership of the Saudi crown prince and prime minister Mohammed bin Salman (who additionally serves as chairman of NEOM's board of directors, when he's not busy overseeing the state's affairs), and backed by a cool $1.5 trillion, this megadevelopment seeks to impress and inspire the wider world with its comprehensive transformation of the country's original landscapes, while simultaneously taking advantage of Saudi Arabia's 'crossroads' location.[13] Saudi Arabia is certainly not wasting any

---

* Reflecting the city's international and avant-garde outlook, the name NEOM is a portmanteau of the Greek *néos* ('new') and the Arabic *mustaqbal* ('future').

time in drawing attention to its unusually connective urban innovation. With the aim of welcoming its first million residents by 2030, earthworks for THE LINE – effectively both NEOM's ligature and its epicentre – commenced in 2021, less than a year after it was first announced.

Suffice to say, THE LINE's placemakers are not short of ambition. One of their main pledges is that the city will be an eco-hub where environmental sustainability is the order of the day.[14] Because there are no roads, there will not be any road bridges, tunnels and street lighting, all of which cost large amounts of money for construction and maintenance, and apply a direct, visible impact on the natural landscape. No roads also means no cars, and therefore no need to cajole people into choosing electric vehicles or taking the bus, or wasting scarce land on parking spaces. As a result, this will be a city devoid of sprawl, a peculiarly connective linear development that champions mobility without encroaching on or damaging the land to either side. Yet most remarkably of all for a nation long associated with oil, Saudi Arabia's flagship city intends to be a zero-carbon society running on 100 per cent renewable energy, primarily wind, solar and green hydrogen from the world's largest plant of its kind.[15] The prevailing aspiration of this self-avowed 'new wonder for the world' is that humans and nature might one day coexist in this postmodern era.[16]

Liveability, a voguish principle among urban planners, research teams and consultants keen on quantifying the basic features of our existence, appears prominently in THE LINE's marketing material as well. Whereas existing cities across the world suffer from all manner of issues associated with rapid population growth – including pollution and traffic congestion, for which both Riyadh and Jeddah are notorious – THE LINE is being designed with efficiency and comfort in mind. Thanks to its uniquely interconnective design, all daily necessities will be available to residents within just a five-minute walk, and immaculate natural landscapes (which, depending on one's specific location, can include deserts, mountains and the coast in merely two. With there being no need to drive along multi-lane motorways from suburban estates, residents will never be stuck in traffic, and will never need to pay for car insurance, petrol or parking, affording them greater time to enjoy the leisure opportunities on offer. Soon enough, they may even forget what life with a car is like: there simply won't be a need to drive one. Instead, residents are encouraged to hit the crystal-clear waters of the Red Sea, where a 'hidden marina' will allow individuals to dock their yacht or jump on a cruise ship safe from hurricanes (take that, Caribbean!) and unimpeded by cargo ships (in your face, Mediterranean!). Unspoiled beaches – one of which will glow in the dark[17] – as well as idyllic islands, alluring desert and snowy mountains are all a quick journey away: if this isn't the good life, what is?

Still not convinced by THE LINE's marketing drive? Perhaps you're concerned about the Arabian Peninsula's summertime temperatures, which frequently exceed 40°C and occasionally reach far higher levels? Careful consideration, THE LINE's placemakers promise, has been given to the size and layout of the city to ensure that sunlight and shade will be perfectly balanced, while NEOM's horticultural oasis, featuring giant greenhouses run according to Dutch intensive farming principles, aspires to create an artificial climate suited to growing crops in this naturally hot, dry region.[18] Perhaps you think that 170 km (105 miles) – approximately the distance from central London to Birmingham – is an awfully long way to commute or to see a friend within the same city? The Spine, the city's ultra-high-speed rail line – which could conceivably be the world's first hyperloop transport system, the brainchild of contemporary society's pre-eminent idealist, Elon Musk – will allow a passenger to travel from end to end in only twenty minutes, a possibility realised by the absence of even a single deceleration-inducing bend.[19] And even better for everyone but trainspotters, because the railway line will be underground, pedestrians enjoying a leisurely stroll will never be disturbed by noisy carriages and unsightly cables. To all intents and purposes, this infrastructure will be imperceptible.

Speaking of aesthetics, aren't the world's most 'liveable' cities – Vancouver, Vienna, Copenhagen, Sydney – typically beautiful places? Even if the railway is underground, surely an imposing, elongated skyscraper would appear rather inconsistent within Arabia's pristine natural landscape? According to the project's designers, there is no reason to worry. With its glass mirrored exterior, THE LINE is being designed to appear effectively invisible, for any onlooker will see the environment behind them as they look forward, with the effect that the human world and natural world seem one and the same. Further, by walking through interconnected green spaces such as parks as well as besides glass-fronted universities, businesses and the like, pedestrians can feel themselves a part of a welcoming, inclusive and thoroughly accessible city, at odds with the dark, narrow passageways and suspicious, concierge-protected institutions of other urban areas. And like all 'liveable' cities, which compete to attract highly educated, skilled workers from across the globe, so too does THE LINE paint itself as the place 'where the best and the brightest live', a laboratory for experimentation, a paragon of luxury living and a haven of world-class preventative healthcare. Anticipating that it will soon compete with the world's most prestigious centres of education and technology – Silicon Valley, Singapore, London, to name but a few – THE LINE's advocates are quick to highlight its own advantages of a warm, dry and sunny climate, an absence of air pollution and, through being designated a special economic zone, investor-friendly taxation laws and governance.[20] It is easy to envisage THE LINE's residents as resembling the contents of a university prospectus: invariably young, fit, active professionals who have come from all corners of the globe to work and experiment together, all with the common goal of creating a better future – while having lots of fun along the way.

Surely every society needs somebody to do the boring jobs, though? This is where robotics and artificial intelligence come in. Instead of requiring people to work in factories or clean toilets, the placemakers claim that all menial and repetitive tasks will be the responsibility of dedicated automatons from the get-go. First, to resolve the daunting issue of constructing an entire city from scratch, AI is being used to pre-engineer the city's various parts, including their electrics, and optimise their production on an industrial scale.[21] Considering that these components are being manufactured locally and will be standardised so that they can be cumulatively tacked on to the city's growing skyscraper block as distinct modules in the manner of a giant Lego set, the expectation is that waste and energy usage (as well as the typical hazards associated with the construction industry) will be minimised, again helping THE LINE realise its sustainability goals.[22] In time, AI will be used to manage the flow of people throughout this 'smart city' as well, for example by recording and thereby learning how many commuters are trying to travel between any two floors at one time, so that the appropriate elevator can be made available to them automatically rather than compelling them to wait. Mooted, too, is the possibility of using AI at the international airport to transport checked bags directly to passengers' destinations, a level of connectivity capable of averting fatigued grumbles at the baggage carousel. Here, the placemakers suggest, are pragmatic solutions to society's most complex – and annoying – problems.

What's more, like any good city, THE LINE promises to continue evolving: it aims to be a dynamic entrepreneurial metropolis where dreamers and innovators can think up new answers to issues if and when they arise. Instead of being commissioned to find solutions for an already inefficient urban system, curtailing their creativity, innovators are encouraged to let their minds run wild, developing prototypes that would seem absurd elsewhere. Already one science fiction mainstay is being trialled: the flying car, or more specifically the Volocopter, a German-manufactured aircraft which, if all goes to plan, will act in place of taxis and emergency vehicles.[23] If constructed as intended, an island theme park with animatronic dinosaurs will allow *Jurassic Park* fans to experience a Middle Eastern Isla Nublar, while an artificial moon will illuminate the city at night.[24] Besides, owing to the city's modular design – consistent with the 'Plug-in City' concept first proposed back in 1964 by one of THE LINE's architects, Peter Cook – should individual modules prove flawed in any way, it will be possible to lift them out, make adjustments and reattach them whenever necessary.[25] In theory, then, no resident will ever need to tolerate an imperfection in their city of discrete and unique but fundamentally interconnected modules.

Part of what is so mesmerising about THE LINE is how it forces us to confront a truth we try to deny: that too many existing cities effectively fail to serve the majority of their residents. When councils slash funding for running libraries and maintaining playgrounds, or cut refuse collection days, but pay to establish or conserve prominent landmarks or make expensive bids to host prestigious events, who is actually benefiting? When they suddenly improve public transport in historically underserved neighbourhoods, are they doing so because they want existing residents to access job opportunities elsewhere in the city, or because they hope these places will now appeal to wealthier people who can gentrify them on the cheap? When new freeways displace tight-knit communities, is anyone or anything meaningfully advantaged? Even parks, which on the surface seem so egalitarian, are typically developed in affluent neighbourhoods, while low-income districts are handed smokestacks and waste disposal sites. Is it not possible to create a place where city and resident are conjoined in a relationship so intimate that the benefits reach everywhere, and everyone?

Alongside growing concern about humans' significant environmental impacts, this ethos has helped engender a burgeoning interest in sustainability among many of the planet's urban planners and policymakers since the 1980s. Although different planners put different weight on the sustainability concept's three main pillars – economic, social and environmental – broadly speaking, so-called sustainable cities are designed so that all three intersect, enabling current and future generations alike to enjoy living in prosperous, healthy and environmentally conscious urban areas.[26] Various cities worldwide have been either planned (Dongtan, China), created (Songdo, South Korea) or modified (Oslo, Norway) according to this principle, often imposing radical yet practical strategies such as closing minor streets to through traffic (Barcelona, Spain), planting vegetation along major roads and establishing new electric bus routes, bike lanes, outdoor escalators and cable cars (Medellín, Colombia), and mandating the installation of green roofs (Basel, Switzerland). Such a trend has not gone unnoticed by the Saudi leadership, with KAUST as well as the Soudah Peaks luxury mountain resort among the new and planned Saudi developments to champion sustainability principles. Where Saudi Arabia does deviate, however, is in its decision to develop a sustainable city according to the simplest and most immediately connective shape.

To be clear, the idea of creating a linear city is far from new. Even disregarding those settlements that have assumed a linear form organically – due to being situated in a narrow mountain valley or along a main road, coastline or river – a loose school of urban planners in a range of countries has advocated lines as means of bringing about various types of social reform since the late nineteenth century. The first model is generally attributed to Arturo Soria y Mata, who in 1882 attempted to apply a linear design to Madrid, organised around transport infrastructure, most notably a loop tramway route that would curve around the city's periphery without needing to pass through the centre first.[27] Championing this design's potential to integrate human needs with the natural environment, Soria suggested that vegetation be planted along streets and green spaces be kept open for residents' enjoyment. Eventually, he hoped, this model would be emulated elsewhere, so that each linear urban system could be connected together directly like the branches of a tree, rather than coalescing randomly as an inelegant sprawl.[28] (The 'garden city' vision of one of Soria's contemporaries, the English urban planner Ebenezer Howard, was somewhat analogous, involving carefully planned clusters of small, interconnected settlements, albeit with radial and orbital rather than linear transport infrastructure.)[29] However, only 5 km (3 miles) of Soria's loop were ever completed, as a combination of financial constraints and, from 1936, the Spanish Civil War, brought the development to a permanent halt.[30] The neighbourhood of Ciudad Lineal has since been absorbed into Madrid's suburbs, although its creator has at least been honoured in the names of a metro station and the main avenue running north-west to south-east.

Subsequent linear city devotees, a broad and generally head-strong group including some of the modern world's most famous architects, have enjoyed even less success. One example is the American Frank Lloyd Wright, who in 1932 unsuccessfully proposed a decentralised settlement called Broadacre City, which would be made up of individual one-acre homesteads connected by a network of superhighways.[31] The Swiss-French Le Corbusier, who infamously proposed razing central Paris's historic Marais neighbourhood in order to build eighteen tower blocks on a rectangular grid, was another prominent architect who enjoyed greater luck with his building designs than his linear city plans.[32] Directly inspired by Soria as well as the human body, for much of his career Le Corbusier was keen to accomplish his *Ville radieuse* ('Radiant City') plan, which would feature high-rise housing blocks stretching in long, straight lines, with pedestrian, cycling, automobile and public transport routes on different levels tying the city together.[33] Although certain of its elements can today be seen in the Indian city of Chandigarh, the *Ville radieuse* generally proved impractical and unpopular; even the young and infamously totalitarian Soviet Union, curious about the possibility of reconstructing Moscow to better reflect its philosophy, rejected the plan.[34] Le Corbusier's later, even more overtly linear proposals were similarly ill-fated. Envisioning a kind of viaduct city comprising fourteen residential levels sitting beneath an elevated highway, his *Plan Obus* ('Shell Plan') for the French Algerian capital of Algiers was never constructed, while his *Cité linéaire industrielle* ('Linear Industrial City'), involving lines of settlement units along key transport arteries in France, remained more theoretical than real.[35]

On paper, the Soviet Union, with its autocratic government and commitment to reconceptualising modern life, would appear to have been the perfect arena for such a radical urban plan. And yet just as Le Corbusier found, other urban planners here struggled to see their linear visions realised. For instance, Nikolai Milyutin proposed that systematically arranged linear cities be incorporated within the country's first Five-Year Plan from 1928, seeing their potential to break down distinctions between the rural and urban proletariat, but was rebuffed on account of their lack of industrial capacity. Although Volgograd (formerly Stalingrad) appears to reflect much of Milyutin's vision, it actually had a linear shape according to a meander of the River Volga even prior to his plan, which was largely imitated when the city was reconstructed following its devastation during the Second World War.[36] Another planner, the German architect Ernst May, did manage to receive official approval to build the new steel city of Magnitogorsk according to a functional, linear design, but his vision was undermined almost immediately by a paucity of material resources and by shifting ideological ambitions.[37] Later, from the mid-1960s, in the Soviet satellite state of Poland, the Finnish-born architect Oskar Hansen advocated parallel belts of urban development running from the Baltic Sea to the Tatra Mountains, but in the end only a few blocks of flats were constructed in Warsaw and Lublin.[38]

For over a hundred years now, linear city plans have been proposed in other countries including the United Kingdom, Germany, Belgium, French Congo (now the Republic of Congo), Japan, Pakistan and Chile, but no matter whether they sound whimsical or pragmatic, they have consistently struggled to gain traction.[39] One might think that these failures would give the Saudi leadership pause, but then again, this is a country whose lexicon does not appear to include a word for epic infrastructural defeat. Despite seeing a number of its own big ideas impeded by a mixture of financial crises, construction delays and lack of public interest – of the six megaprojects announced in 2005, only King Abdullah Economic City was ever launched, and it remains home to just a few thousand people[40] – Saudi Arabia shows no signs of capitulating. Given the challenges the earth – and Saudi Arabia specifically – are projected to face over the coming decades, the country recognises that responses must be at least as pragmatic as they are sensational, and possesses the resolve as well as the prerequisite funds to continue experimenting. Between 2023 and 2050, the planet's urban population is forecast to more than double.[41] With an urban-based and youthful population – 85 per cent of its 36 million inhabitants live in urban areas, and almost half are under the age of twenty-five[42] – Saudi Arabia and its straining cities risk demographic disaster. Coupled with the difficulties posed by climate change for a hot, dry country already short on arable land and freshwater sources, proactivity here is crucial. Suddenly, the idea of building a necessarily huge city on the smallest territory possible makes a lot of sense. Perhaps this is the solution to all the country's problems.

Just as importantly, instead of feeling encumbered by the challenges of raising and managing its sizeable youthful population, Saudi Arabia is viewing it as an asset. A major problem with many cities today is that expensive downtown rents and a lack of housing stock compel many young adults to live far from places of work and leisure, so why not flip the script and integrate all three: why not have two buildings constituting an entire city, where living spaces are fully connected within rather than isolated from the wider urban fabric? No more 'affordable' and social housing tower blocks situated far from most people's workplaces, in line with Le Corbusier's vision. No more business districts from which commuters depart in the evening and don't visit at all during the weekend, leaving them empty and devoid of life. THE LINE's interconnective design may allow human energy to flow incessantly, for the dual skyscrapers will always be functioning. Supplemented by the natural vibrancy young people are presumed to bring, it is plausible that this youth-oriented city will be able to attract newcomers and avoid the failings of previous linear urban developments, while exerting the smallest possible impact on the natural environment.

Theoretically, THE LINE is incomparable in its potential to reshape urban life and humans' engagement with the planet. Backed by near-limitless funds and bringing together internationally renowned urban planners and architects, it is a flabbergasting collaborative exercise that promises to draw onlookers' attention to a part of the planet that despite – or perhaps because of – its placement towards the heart of many world maps, has for centuries been reduced to a blurry contact zone between a better defined 'West' and 'East'. By pledging to circumvent the greatest problems cities pose to the natural world, while offering a lifestyle it hopes will appeal to professionals of different stripes, it represents a novel model of urban development that should interest people globally.

However, the very notion of constructing an entire city from scratch, designed to connect disparate places as two continuous skyscrapers, also arouses a number of important concerns. In spite of the project's purported sustainability and protection of the natural world, THE LINE will inevitably have an environmental impact – and one far greater than the city's marketing material would have readers believe. The construction phase will involve the development of staggering amounts of glass, steel and concrete, necessarily entailing a (rather sizeable) carbon footprint.[43] Even once the city is complete, it will – barring a radical evolution in air travel in the interim – produce carbon emissions from its international airport, while the use of cloud seeding to create artificial rain may spread toxic substances such as silver iodide.[44]

Wind and solar are both logical sources of energy in this region, but the sheer amount of energy required to power two buildings of this height and length will doubtless be daunting, and it is not out of the question that in time, the placemakers will discard some of their more principled intentions. Informative in this regard is the fate of Masdar City in the United Arab Emirates, a settlement originally conceived in accordance with various sustainable principles, including walkability, carbon neutrality, zero waste and a prohibition on private vehicles (yet ironically opening in the shadow of Abu Dhabi International Airport), but which has since abandoned many of these precepts, becoming a 'low-' rather than 'zero-'carbon development with significantly downgraded water and waste management objectives and, even more ominously, a population ten times smaller than the 50,000 residents projected.[45] An additional issue – and one directly relevant to THE LINE's distinctive design – is that without the development of some kind of sophisticated eco-corridor, a city of this sort will inevitably act as a physical barrier to migrating and pollinating species, as opposed to assimilating seamlessly into its surrounding natural landscapes.[46] Flying animals in particular are vulnerable to this development, for the simple reason that because its exterior walls will be mirrored, they are at great risk of unwittingly crashing into the city.[47] Ultimately, in all the noise about how this city will represent something completely novel, an important truism seems to have been forgotten: nowhere is a completely blank slate. As loudly as one may assert that the city will constitute an 'environmental solution to urbanism',[48] in reality, any urban area is bound to affect and impair the natural environment in one way or another. Its configuration, in this respect, is immaterial.

In fact, THE LINE's singularly connective design is also bound to breed distinctive practical problems of its own. First of all, consider that a line is the shape with the greatest feasible distance between its two furthest points, and no unquestioned centre. Unlike their conventional circular-city counterparts, who can reach a mutually agreed site quite easily from any point on the circumference, two people living at either end of a linear city are therefore kept as far away from each other as possible. Sure, they can hop on the ultra-high-speed train connecting each and every part of the city, but if all residents were to live just minutes from a station on the same route, then an awful lot of stations – one study estimates eighty-six – will be needed.[49] Suddenly the twenty-minute rail journey from end to end appears a gross misrepresentation of life along THE LINE, being true only of the four-station Spine. Most people will instead take far slower services, plausibly with an average commute time of an hour at least, before we even consider the possibility of train delays due to signal malfunctioning or station failures. One break in the link tying everything together could effectively paralyse a large portion of the city, cutting off millions of people without there being a single suitable alternative to temporarily take the railway's place. Indeed, as we have already seen in the cases of Mozambique and Panama, building connecting routes risks simultaneously building dependence, and so where no effective substitute exists, the entire functioning of human society is liable to collapse.

Plus, even if THE LINE is constructed as planned, there is cause for scepticism that its perfectly linear shape will be maintained for ever. Illuminating in this respect is the fate of the most successful previous linear city plan, which was realised on a large scale, but preserved for only a limited time. The winning candidate of a 1957 competition to design a new, centrally located capital for Brazil, Lúcio Costa's concept for Brasília was, essentially, a cross formed by the meeting of two lines, running north–south and west–east. Though the area's natural topography demanded that one of the axes be bent into a boomerang shape, the final plan proved to be more befitting of the government's forward- and outward-facing ideals, bearing an obvious likeness to the quintessential symbol of jet age connectivity and one adored by Costa's hero Le Corbusier: an aeroplane.[50] To this day, the city's two principal axes – one straight like a fuselage, the other curving like a pair of wings – are easy to spot from above. However, because Brasília's population is now far bigger than initially envisioned, new solutions have been required on the fly.[51] Disorganised sprawl spreads to the west of the city, creating new centres of activity away from the notional cross, as residents choose how they want to engage with their city (or are forced to live in previously undeveloped areas because they are unable to afford a place in the original city), rather than adhering to a planner's ideals.[52] The lesson that THE LINE's placemakers may want to take away from Brasília's fate is that no matter how lucid one's vision, cities typically grow rather organically in time, as not all aspects of people's engagement with urban life can be fully regulated. Their retort – more likely in private than in public – is that they have the 'advantage' of an absolute leader willing to accomplish and perpetuate this imaginative vision no matter what.

After all, implicit in the city's unique design, which necessarily obligates residents to live their lives a certain way, is a broader question about control. Saudi Arabia's current programme of investing heavily in AI and megadevelopments as well as offering new subsidies for education and housing looks far more like paternalism than altruism when one considers that this strategy was prompted by a fear of revolution from 2010 to 2011, when social unrest sprang forth across most of the Arab nations of North Africa and the Middle East (in the process creating power vacuums that Saudi Arabia and its Shia-dominated rival Iran still vie to fill).[53] For this reason, THE LINE has been conceived not merely as a technology hub, but as a tool for the Saudi leadership to buy its survival in the long run. Why else would one build a city that comprises two effectively transparent buildings? With 24/7 surveillance and AI technology present at every turn, residents' lives will be intimately regulated and constantly monitored. Every journey, every visit to the opera, every *action* will be inspected, the statistics being fed into a citywide database. Yes, the data may help AI to make residents' lives more efficient and less frustrating, but in a country whose ruler has a reputation for callously shutting down dissent, at what cost? Given the hoo-ha raised by the possibility of one day having to carry national identification cards in countries including the United Kingdom, United States and Australia, on top of even broader criticisms that so much of modern life is marked by surveillance – CCTV, website cookies, drones, licence-plate cameras, electronic toll roads, fast-pass lanes – life on THE LINE, with its pervasive drone and facial-recognition technology,[54] seems unlikely to appeal to many people who value individual freedoms. A liberal society, one must remember, this is not. Further, though in theory certain adjustments can be made to individual modules based on residents' engagement with the city, so far the project has been administered fully from the top down, with no input requested from potential residents.* Given that it is unlikely that this type of control will be lifted even once the city is opened, prospective citizens will need to decide carefully whether they are willing to adhere to a very specific set of norms before moving in.

---

* A related concern has already been aired by one of THE LINE's original architects and now a critic of the project, Wolf Prix, who contends that two linear skyscrapers will appeal only to young adults. The fact that these megastructures are still forming the basis of the city would imply that his apprehensions are not held by the placemaking team.

Herein lies the other challenge facing THE LINE, and Saudi Arabia more generally. As a city deeply reliant on international cooperation for its conceptualisation, construction and eventual habitation, cultivating interest and appeal to bring people from all corners of the globe to a single nexus is paramount. However, with popularity and renown comes an increasingly intense media spotlight, capable of dissuading migration to Saudi Arabia and possibly undermining the success of the project before it is even completed. In the first instance, just as the 2022 FIFA men's World Cup host Qatar attracted widespread attention for its use of *kafala*, a migrant worker sponsorship system that gives employers near-total control over their labourers' lives (including potentially confiscating their passports, withholding their pay and coercing them to work long hours on dangerous construction projects with minimal risk of facing legal repercussions), so too is Saudi Arabia facing growing international scrutiny for numerous cases of worker exploitation on megaproject developments.[55] Although it announced the reform of its *kafala* system in March 2021 to allow some migrant workers to switch jobs and leave the country without their employer's consent under a limited set of circumstances, labourers remain almost completely dependent on their sponsors, with no possibility of unionising or appealing against abuses.[56]

Meanwhile, there is significant reason for concern that certain communities are being effectively eradicated from the country's future vision. In 2017, Saudi security forces ravaged Awamiya in the Eastern Province, a historic city with a predominantly Shia Muslim population, whose political dissidents and militants had come to be seen as a serious security threat to the country's leadership.[57] Since late 2021, hundreds of thousands of foreign nationals have been expelled from Jeddah without compensation in the name of 'revitalising' a city that seeks to become a world-class destination over the next decade.[58] THE LINE has been embroiled in similar controversy. As many as 20,000 Indigenous Huwaitat people living along the city's future course – across much of an area bin Salman has described as 'untouched' and 'almost empty'[59] – have been forcibly evicted since 2020. Dozens of them have been threatened, arrested and detained for resisting.[60] Several have even been sentenced to death, while another, Abdul Rahim al-Huwaiti, was shot dead by security forces at his home in Tabuk province for posting videos to social media in an effort to raise awareness of his people's plight. As this sorry case shows, in its determination to revamp its public image from camels in the desert to luxury living, Saudi Arabia is far less interested in large segments of its population than it is in showcasing itself to a privileged international audience.

With the world's gaze now fixed on Saudi Arabia, its slick branding campaign is thus a form of window dressing, a vanity project meticulously designed to beguile and divert attention from the country's dire human rights record as much as it is about re-inventing the city. With inferior individual freedoms to the likes of Iran and Libya according to the research institute Freedom House, but far superior economic prosperity, Saudi Arabia is taking advantage of the oil money that made it rich in order to coax international interest and investment.[61] Although human rights organisations have put pressure on NEOM's foreign partners to reconsider their involvement,[62] most remain faithful to the project, arguing that the opportunity to create a lasting conceptual and technological legacy in this new boom region is too tantalising to pass up. An exception is the British architect Norman Foster, known for structures such as 30 St Mary Axe ('the Gherkin') in London, the HSBC Building in Hong Kong and the renovated Reichstag in Berlin, who quickly terminated his involvement on NEOM's advisory board following the suspected state-sponsored murder of the US-based Saudi journalist Jamal Khashoggi in 2018.[63] But with most others choosing to turn a blind eye to the Saudi leadership's merciless treatment of anyone who challenges its monopoly – whether consciously as activists or militants, or through merely existing in the case of the country's clandestine LGBTQIA+ community – THE LINE is an urban embodiment of magpie syndrome, legitimising the regime's more dubious practices by disguising them behind a veneer of technological and social advancement.

Certainly, despite Saudi Arabia's efforts to reinvent the world's principal social unit and in the process overhaul its own international image, it's important to keep in mind that THE LINE's primary purpose has less to do with the sustainable, liveable credo repeated by the city's placemakers than with leveraging the country's unique combination of location, space, money and autocracy. After all, if this really was a project aimed at inspiring a new model of urban development to be imitated elsewhere, then surely it would be a lot more practicable than an eye-wateringly expensive construction requiring abundant open space and a casual disregard for human livelihoods, all overseen by an unimpeded government. One impossible-to-replicate linear city hardly entails a paradigm shift. Instead, THE LINE represents Saudi Arabia's principal means of creating a Saudi-centric world, a distinctly interconnected and unprecedented phantasmagoria where respected specialists from across the globe can innovate to their hearts' content, unencumbered by medieval street networks and industrial-era homes. Though construction is already far behind schedule – in 2024, the Saudi government felt compelled to admit that less than 2.5 km (just 1½ miles) would be ready by 2030,[64] hardly ideal considering the city's sportswashing potential as a host of the 2034 men's World Cup – the sheer audacity of THE LINE should allow it to have much of the impact its government desires regardless. In time, the hope is that curious eyes, transfixed by wonder, will come to overlook Saudi Arabia's more archaic laws and its leaders' predilection for discipline and control, and instead see the country as a logical command centre for a rapidly changing planet.

In a way, THE LINE's unconventional design is a perfect

metaphor for the city itself. The two elongated skyscrapers promise to offer residents unparalleled connectivity from one end of the city to the other (and from there, they can travel easily to other global hubs), yet for those on the outside, they represent the formidable walls of an exclusive dreamworld. THE LINE's mirrored facade could offer a window into a future in which artificial intelligence is not to be feared (as *Terminator*, *Ex Machina* or *2001: A Space Odyssey* might suggest), but rather enhances our own existence, and in which brilliant people finally manage to bring the natural and human worlds together. However, and most compellingly of all, the same exterior also implies that ultimately, we see what we want to see. For some, THE LINE is central to 'creating the future of the human race'.[65] For others, it is bringing Paolo Bacigalupi's best-selling novel *The Water Knife*, whose setting is a desert utopia for the elite that instigates dystopia for everyone else, to life.[66] It is very possible that THE LINE will suffer a similarly anticlimactic fate as Pyongyang's monumental 'Hotel of Doom' or Malaysia's $100 billion Forest City 'ghost town', and yet it almost doesn't matter how THE LINE actually evolves as a self-contained place. In a globalised world which demands that we contemplate proceedings and crave innovations beyond our country's borders, geographical connections are a powerful asset. Whatever our hopes and aspirations for the future of humanity – cities that are more sustainable, societies that are more equitable, environments that are preserved rather than destroyed – the very *idea* of THE LINE exemplifies how we ignore connectivity at our peril. Whatever our personal assessment of the city, and the country it aims to position at the planet's conceptual nucleus, THE LINE serves a purpose as long as it prompts us to look at the world in a new light, a place where everything must be interlinked, whether we like it or not.

5

Resistance: The Baltic Way

The Baltic Way is the only way to freedom,
brotherhood and equality on the shores of our
common Baltic Sea ... We are ready, we are arriving.

Joint statement by Rahvarinne, Tautas fronte and Sąjūdis[1]

There are some moments in history so momentous, so powerful, so *photogenic*, that the snapshots that remain of them have achieved iconic status in the cultural consciousness. Shopping bags in hand, a mysterious man in a white shirt stands in front of a column of tanks departing Beijing's Tiananmen Square. A procession of plain *khadi**-wearing men in India follows its scrawny leader to the coast. In Saigon, now officially known as Ho Chi Minh City, a Vietnamese Buddhist monk sets himself on fire on the street. Two African American athletes raise a black-gloved fist from the Olympic medal podium in Mexico City. A group of women in red cloaks and white bonnets walks two by two outside the US Capitol, beneath the steps up which sixty people with physical disabilities had dragged themselves with difficulty decades earlier. Outside the Swedish parliament, a teenaged girl sits with a homemade sign. Its bold message, *Skolstrejk för klimatet* ('School strike for the climate'), is as eye-catching as the oversized yellow jacket she wears.

So many of the most compelling forms of modern civil disobedience are profoundly photogenic, capable of instantly capturing the attention and intrigue of viewers worldwide. Whereas some protesters seek to highlight their cause quite viscerally, for instance by juxtaposing bright colours and provocative language with solemn locations – the Russian punk rockers Pussy Riot's choice of neon balaclavas and expletive-laden lyrics in Moscow's Cathedral of Christ the Saviour being an excellent example

---

* A hand-spun and woven cloth, which Mahatma Gandhi promoted as part of his Swadeshi (self-sufficiency) movement.

– others prefer to accentuate more banal items, such as by wielding umbrellas en masse as protection against pepper spray (Hong Kong), raising blank white sheets of paper to symbolise censorship (China) or towing suitcases to represent the travel required to obtain an abortion in another territory (Northern Ireland). Nonetheless, few examples of peaceful protest have ever been as seemingly rudimentary and yet as ambitious and impactful as what might reasonably be described as the world's first flash mob. Combining the visual intensity of scale with the simultaneous ordinariness and incongruousness of 'regular people', punk and priest standing side by side, the Baltic Way* was original in utilising geographical connection as a form of protest capable of bringing down a superpower. In doing so, it changed the world.

It is no exaggeration to say that the contemporary map of Eastern Europe would look very different without this demonstration. For half a century, the deep red of the Soviet Union reduced a chain of nations to a mere frontier region, a protective barrier against Western decadence and injustice. Seemingly forgotten by millions across the 'Iron Curtain', the three northernmost nations in this buffer risked being consigned to the history books in the manner of Prussia, Burgundy or Novgorod. However, in becoming more connected to each other than to the confederation as a whole, they instead proved to be its destroyer. Though it existed only momentarily as a physical entity, the human chain joining Tallinn to Vilnius via Rīga, the capitals of Estonia, Lithuania and Latvia respectively, manifested a popular and enduring desire to look westwards, to turn one's back on the Russian Bear while sharing a bond as ideological as it was geographical. Not merely an idealistic statement of each nation's sovereignty and

---

* Alternatively called the 'Baltic Chain'. The linguistic variation is due to the choice of words in the respective languages: whereas the Estonian word *kett* means 'chain', the Latvian and Lithuanian words, *ceļš* and *kelias*, respectively, suggest a 'way' or 'road'.

independence, the Baltic Way provided a tangible statement of where the Soviet Union's power and influence ended, no matter what a treaty might imply, or what maps displaying its borders might suggest.

The extraordinariness of this protest lay partly in its coordination, and partly in its unique ability to tie the momentous with the truly *ordinary*: historic castles were connected with everyday villages; popular thoroughfares and back-country lanes were joined together as one. Yet no less remarkably, it fastened together three nations that are far more different than many international observers realise. Whereas Estonia and Latvia share a long Germanic history – both were conquered by Teutonic Crusaders and their coastal towns were integral to the Hanseatic League, a medieval confederation of merchant guilds and market towns which essentially represented an urban-based precursor to the contemporary European Union – the multi-ethnic Grand Duchy of Lithuania was at one point Europe's largest state[2] as well as the last to retain paganism as its state religion.* By the mid-seventeenth century, more or less the entirety of Estonia and much of Latvia were part of the Swedish Empire, whereas Lithuania had long been united with Poland and remained one half of an independent commonwealth (with authority over the rest of modern-day Latvia) until the late eighteenth century, when it was incrementally carved up by its Russian and Prussian neighbours. Religion also distinguishes the countries: most Lithuanians today

---

* The Teutonic Order enjoyed little success in its efforts to invade and forcibly Christianise powerful Lithuania: even after Grand Duke Jogaila converted his country to Catholicism on marrying Queen Jadwiga of Poland in 1386, territory-hungry Crusaders continued to raid the new Polish–Lithuanian Commonwealth until their decisive defeat at Grunwald in 1410, in one of medieval Europe's greatest battles. Meanwhile, Hanseatic merchants struggled to gain a foothold even in the thin Baltic Sea sliver of Lithuania around Memel (today Klaipėda), as the Grand Duchy tended to be suspicious of the rival League's territorial ambitions.

share with Poles a preference for Roman Catholicism, whereas Latvia and Estonia are home to larger Lutheran and Eastern Orthodox communities than their southern counterpart, and around 60 per cent of Estonians declare having no religion at all. And linguistically, Estonian is far closer to Finnish and Hungarian than it is to Latvian and Lithuanian, which bear some rather surprising similarities to the classical Indian language of Sanskrit.[3]

Given these stark differences, it is unsurprising that the three countries vary in their modern international outlook as well. Notwithstanding Estonia and Latvia's particularly large ethnic Russian minorities, some of whom look eastwards first and foremost,[4] Latvia and Lithuania typically prioritise their shared relationship, viewing each other affectionately as *braliukas*: 'little brothers'. By contrast, Estonia has long yearned to be treated as a fully fledged Nordic nation,[5] connected conceptually and emotionally to Finland in particular.* In line with viewing itself as quite different from the nations to the south, since achieving independence from the Soviet Union in 1991 Estonia has indefatigably sought to become one of the world's most profoundly connected societies, an 'e-Estonia' where internet access was officially declared a human right as early as 2000, where all schools enjoyed an internet connection by 2001 and where voters could elect officials electronically before any other country (2005 for local elections, 2007 for parliamentary elections).[6] Home to Skype[†] and since 2014 offering e-residency to people seeking to run a European Union-based company online wherever they are

---

* Consider that the Estonian national epic poem *Kalevipoeg* is very closely related to its Finnish counterpart the *Kalevala*, the latter of which popularised the mythical place Pohjola ('northern lands'). Many in Estonia still seek to reinforce their nation's relationship with Finland's imaginative geography: the name Põhjala appears in a range of contexts, including the country's largest craft brewery.

† Although its co-founders were Swedish and Danish, Skype was developed by Estonians and launched in Tallinn.

in the world, far from Estonia being seen as 'just another' post-Soviet Eastern European nation, digital connections have helped the country transcend borders and more clearly define its place on the metaphorical map.

What, then, do Estonia, Latvia and Lithuania share, which would help them cooperate in their moment of internationally televised civil disobedience? First, a proud choral tradition. All three countries have hosted large-scale song festivals since the nineteenth century, bringing together tens of thousands of ordinary citizens keen to sing (and in some cases dance), with an emphasis on a cappella. These festivals can be among the largest choral events anywhere in the world in a given year, and are today recognised by UNESCO as examples of intangible cultural heritage.[7] UNESCO additionally acknowledges Lithuania's *sutartinės*, multi-part folk songs historically performed by women in the country's north-east,[8] while in all three nations even the most reticent of vocalists will engage in a midsummer singalong after a few beers. This mutual heritage of collective music-making acts as an important social glue, a harmonious, polyphonic symbol of cultural identity that allows each nation to view itself as distinct, but whose very essence implores others to join in. Here lies the second commonality. Even if for centuries most people in Estonia engaged rarely if ever with their counterparts in Lithuania, let alone recognised a shared regional identity or experience, they and Latvians found seeds of unity and unprecedented collaboration, counterintuitively – from the USSR's perspective – in Soviet divide and rule. For in facing the same existential threat to their varying senses of cultural distinctiveness, partnership became a possibility – and a necessity. No longer would the three nations consider themselves detached. Now was the time to combine as a single force, reclaiming their individual sovereignties and identities and restoring them in minds and on maps, while creating a categorically Baltic sense of self.[9]

*

While the concept of a long chain of people holding hands could hardly have been more wholesome as an unequivocal statement of Baltic unity, the anniversary it marked was grounded on secrecy and betrayal. Co-signed on 23 August 1939 by the foreign ministers Joachim von Ribbentrop and Vyacheslav Molotov, the non-aggression treaty between Nazi Germany and the USSR contained a clandestine protocol dividing much of Eastern Europe into separate spheres of influence. A week later, Germany invaded Poland – which the two countries had bilaterally agreed to partition – initiating the Second World War. The USSR followed suit from the east not long afterwards, enabling it to establish a new border with Germany running through Poland's heart. By the middle of 1940, Estonia, Latvia and Lithuania* as well as Finland and Romania's Bessarabia and Northern Bukovina regions were all occupied by the Soviet Union, tying these historic nations together as a continuous geographical strip from north to south marking the western frontier of Joseph Stalin's geostrategic domain.[10] Meanwhile, Molotov lent his name to a variety of improvised, hand-thrown firebomb – although not intentionally, for the term was coined by Finns who, in a sardonic rejoinder to the Soviet foreign minister's allegation that his tanks were merely supplying food, decided that a good meal needs a bit of spice.[11]

Even after the details of the Molotov–Ribbentrop Pact were finally revealed to the public during the Nuremberg Trials (1945–6), at which Nazi leaders were charged with war crimes and crimes against humanity, the Soviet leadership denied the existence of any agreement.[12] More brazenly still, it claimed that Estonia, Latvia and Lithuania had all asked to join the USSR voluntarily in an act of self-determination – even though citizens of all three had attempted to restore their independence after Germany's

---

* Lithuania had actually been assigned to the German sphere, but a second secret protocol in September 1939 'legitimised' a Soviet invasion instead.

surprising invasion of the Soviet Union in 1941* – and therefore it was doing them a favour.[13] However, though this alternative version of history held obvious appeal to the most chauvinistic of Soviets, the reality was less readily forgotten in the Baltic.

For over four decades, the USSR consciously executed a comprehensive process of sovietisation in the Baltic states, designed to forcibly assimilate this geopolitical region into the communist sphere by undermining its populations' distinctive national and cultural identities and curbing citizens' freedom of expression. From 14 June 1941, a date still observed in these countries as a day of mourning, hundreds of thousands of dissidents were deported to Gulag labour camps in some of the bleakest corners of the country, where many were worked to their death.[14] Meanwhile, the significantly diminished population remaining in the Baltic states was compelled to adapt to an unprecedentedly dogmatic society, where manufacturing took precedence and the rural economy progressively collectivised even as it fell into obvious decline.[15] Schools became vehicles of *re*-education, in which teachers had to discourage students from wearing religious symbols or the colours of the national flag, publicly renounce past actions as

---

* Incumbent Russian president Vladimir Putin's attempts to justify his invasion of Ukraine by alleging that the latter is full of Nazis echo the Soviet Union's charge that anyone who opposed its occupation of the Baltic states was pro-Nazi. The reality was far more complex. Estonians, Latvians and Lithuanians fought in both the Russian and German armies, and the vast majority were conscripted by force, though some collaborated with the Nazis because they expected them to liberate their nations from Soviet occupation. Reflecting the complications this period engendered for their historical narratives, Latvia and Lithuania (but not Estonia) today forbid the display of Soviet and Nazi symbols at public events, yet memorials commemorating Nazi collaborators still exist in all three Baltic states. Meanwhile, ongoing Russian aggression in Ukraine has prompted all three Baltic states to tear down their remaining Soviet memorials (including Rīga's Victory Monument, at 80 metres/260 feet one of Latvia's tallest structures) because these are widely regarded as glorifying Russian imperialism, a perspective not held by all.

inconsequential as travelling abroad to learn a new language, and depict their nation's history as, somehow, simultaneously backward and bourgeois.[16] Traditional choral festivals were heavily scrutinised to ensure that no song referred to sovereignty or independence, even if some canny lyricists managed to evade the censors by subtly alluding to locally relevant tales unfamiliar to the authorities.[17]

While the post-Stalin era did also see the revival of some beloved Baltic oeuvres, in time the Soviet leadership opted to fortify its old approach once again, expanding usage of the Russian language in administration and mass communication, and mandating that it be taught in school.[18] Rebellion risked grave and lasting repercussions, as Latvia found to its cost in the late 1950s, when it refused to give the Russian language primacy in its schools, demanded that government employees demonstrate at least conversational proficiency in Latvian, and created a residency permit system that covertly discriminated against ethnic Russians.[19] Although the strategy was briefly effective in preserving Latvian language and culture in Rīga, it soon backfired badly: Latvia's nationalist politicians were duly purged, and replaced by ethnic Russians and Russified Latvians, enticed by new privileges in housing as well as employment.[20] As a consequence of this demographic engineering, which effectively caused Latvia to become more Russified than its Baltic neighbours (though Estonia was also impacted significantly by Russian immigration), the share of ethnic Latvians in its population declined from three-quarters prior to the Second World War to about half by the end of the Soviet occupation.[21] With the Baltic states concealed on maps and in textbooks by Russian labels and unmarked borders, the possibility of their being forgotten by younger generations was real.

However, the social and political landscape of the Baltics and the Soviet Union as a whole was suddenly transformed in the mid-1980s with the implementation of two political reforms by the country's leadership. While perestroika ('restructuring') sought to democratise the political system and introduced some features of a free-market economy, glasnost ('openness') required state institutions to become more transparent in their activities and allowed the media to disseminate news with far fewer restrictions than before. Though these reforms primarily aimed to benefit the USSR as a confederation, for they were regarded as a necessary but belated means of overcoming a long and embarrassing period of economic stagnation, they inadvertently provided additional preconditions for the country's demise.[22] Given a modicum of free speech, media figures started pushing the boundaries of Soviet tolerance.[23] In the Baltic region, journalists found themselves able to promote distinctly nationalistic sentiments and publicise upcoming demonstrations with little to no censorship.[24] Conferences held by new nationalist movements were frequently broadcast live over radio and reported widely on television.[25] Programmes in Baltic residents' vernacular languages became increasingly common, and although nationalist newspapers tended to be regarded as less filtered and therefore more trustworthy than television, both communication channels offered new opportunities to disseminate political information.[26]

Importantly in this respect, Baltic residents were increasingly able to see and hear perspectives from foreign figures regarding their plight. In 1986, television news in Latvia alluded to a remark made by Jack Matlock, a senior adviser to the United States president, that his country never recognised the Baltic states' forcible annexation as legitimate.[27] Baltic viewers' assurance that they were not alone in the fight for freedom was further intensified through international media reports showing strangers in the 'Western' world protesting peacefully against totalitarianism, specifically the Soviet regime. Initiated in 1986 by Central and Eastern European émigrés in Canada, these 'Black Ribbon' demonstrations were coordinated to occur on 23 August, the day the Molotov–Ribbentrop Pact had been signed nearly half a century earlier.[28] Inspired by the demonstrations and emboldened by a successful flower laying at Rīga's Freedom Monument* two months earlier in commemoration of the many Latvians deported to the Gulag, in 1987 the Baltic capitals each held their own Black Ribbon Day protests, many participants wearing black ribbons pinned to their chests to publicly mourn lost relatives and friends and ensure that Soviet injustices in the Baltic region would not be forgotten. Though others stayed home, fearful of reprisal by the authorities – and arrests were made in Rīga and Vilnius – the following year, the Soviet authorities agreed to sanction equivalent demonstrations.[29] For nationalist leaders in the Baltics the message was now clear: in a rapidly changing era, the USSR's iron fist was loosening its grip.

Quickly, protest became a frequent occurrence, with a distinctly Baltic tinge, as Estonia, Latvia and Lithuania's long choral traditions were revived as political actions. Massive festivals enabled people from all walks of life to convene and sing songs whose lyrics were patriotic enough to animate the more nationalistic, but cryptic enough to engage those more loath to publicly challenge the Soviet regime. Even beyond popular venues such as the Tallinn Song Festival Grounds, where some estimate (though likely exaggerate) that 300,000 Estonians descended on 11 September 1988 to sing long into the night, unstylised, improvised folk performances were now commonplace in the streets of Baltic towns, in a clear departure from the staged, carefully censored music and dance promoted by the Soviet leadership.[30] With lyrics in Estonian, Latvian and Lithuanian recounting how three fatigued sisters had summoned up the energy to defend their mutual honour, one song in particular came to symbolise the emergence of a newly pan-Baltic identity, a necessary cry for cooperation between a trio of nations roused to take control over

---

* Constructed thanks to public donations, the monument was unveiled in 1935 to honour Latvian soldiers killed during the nation's war of independence from the nascent Soviet Union a decade and a half earlier.

their intertwined futures.* After decades of nationalist hibernation, residents of the three countries were roused by the sounds of this 'Singing Revolution', and finally felt proud and able to describe themselves as Estonian, Latvian or Lithuanian once again.

Invigorated by the growth of new nationalist groups – the Baltic states' first non-communist political movements since their annexation – in early 1989, attention rapidly turned from securing increased autonomy within the USSR to achieving full independence.[31] Crucially, instead of pursuing independence individually, their leaders recognised that the prospects for any and therefore all three hinged on successful collaboration.[32] Historical and cultural differences would be irrelevant in this endeavour: what mattered was that they shared a mutual enemy, and a preference for peaceful protest over violent resistance. Combining representatives from Estonia, Latvia and Lithuania's most prominent nationalist organisations, Rahvarinne, Tautas fronte and Sąjūdis (the Popular Front of Estonia, the Popular Front of Latvia and the Reform Movement of Lithuania, respectively), the new Baltic Council reflected this growing feeling that the three nations were tied together emotionally and ideologically. And in almost no time, discussion turned to the possibility of making the three nations' abstract connectivity more tangible, even if only temporarily.

Though it's possible the concept of a human chain had first been floated by Rahvarinne's co-founder and leader Edgar Savisaar† a month earlier,[33] we know that the Baltic Council agreed

---

* Written by the Latvian rock musician Boriss Rezņiks and titled *Ärgake, Baltimaad* in Estonian, *Atmostas Baltija* in Latvian and *Bunda jau Baltija* in Lithuanian, 'The Baltics Are Waking Up' became the primary anthem of the Baltic Way.
† Savisaar's political career was rarely tranquil. Having already served as independent Estonia's interim prime minister, he became minister of the interior in 1995, but was soon forced to resign following a secret tape scandal. In 2015 he was suspended from his role as mayor of Tallinn during

to execute the plan while convening in Pärnu, Estonia on 15 July 1989. Through physically joining people in all three nations, such a linear protest would not only provide an explicit declaration of Baltic unity, but also operate as something of a public referendum on the future of their nations. A month later, in a secluded location near Cēsis, Latvia – for in another sign of the nations' strengthening bonds, council meetings would rotate between the three – activists agreed to put the plan into action and began mapping out a route connecting their capital cities.[34]

Part of the decision-making process centred on the endpoints of the chain. If this undertaking were to be successful in conveying to the Soviet leadership a popular demand for Baltic independence, then the sites at either end would need to be associated unambiguously with distinct national identities. The choices were logical and coherent. In the north, Toompea Castle in Tallinn, a majestic complex remoulded since the thirteenth century by various rulers, which combines fortress, palace and parliament and, by sitting at the top of a steep hill of the same name, represents a prominent feature of the city's skyline. In the south, Gediminas' Tower in Vilnius, the remaining part of the city's fifteenth-century hilltop brick castle complex which had been constructed under the leadership of Vytautas, Lithuania's greatest national hero, who expanded the Grand Duchy's territory further than any other leader.

Both towers were, and still are, powerful symbols of their respective nations' military and political histories, and having each withstood several centuries of history, they provide a strong degree of consistency amid the Baltic states' many changes and traumas. To connect the two via Rīga's Freedom Monument, the

---

a corruption investigation, and although he was eventually released from criminal proceedings on account of his declining health, his reputation never really recovered, especially due to his Centre Party's formal ties to – ironically for an Estonian independence leader – Vladimir Putin. Savisaar died in 2022.

Baltic Way's organisers calculated that they would need at least 400,000 individuals, assuming that each person would stand with arms outstretched. Yet they ambitiously added some 80 km (over 50 miles) to the route to increase access for residents of eastern Estonia and Latvia. As a consequence, instead of following the principal road, the Via Baltica, which runs parallel to the Gulf of Rīga's shoreline, the route would take a detour inland through these nations' hearts before reaching the Latvian capital.[35] Measuring nearly 700 km (430 miles), this human chain would be the longest of its kind in history* and, unlike more recent analogues from across the world, which have tended to be limited to self-contained regions, it would overstep national boundaries to manifest a shared and distinctly *Baltic* identity.[36]

One last important question then remained: when should this protest be held? To the organisers, only one date was appropriate for such a momentous occasion – even though they would have just a week and a half to prepare: 23 August 1989, the fiftieth anniversary of the Molotov–Ribbentrop Pact.[37] As was to be expected, from this point on, the planning was frantic. While the three nations assumed slightly different strategies, they all opted to divide their respective sections of the route into discrete portions, which would be overseen by individual volunteers. Local newspapers across each country published information

---

* This record has since changed hands various times, and today stands at over 18,000 km/11,000 miles (in Bihar, India in 2020, when more than 50 million people held hands to increase awareness about a variety of environmental and social issues). A previous effort to create an even more extensive chain took place in 1986, when more than 5 million people participated in 'Hands Across America', a public fundraising event with the goal of creating a human chain across the contiguous United States. The chain was discontinuous – the sheer distances involved as well as the presence of harsh environments such as the desert rendered the objective overly ambitious – but its outcome was laudable nonetheless: it raised an estimated $15 million for the fight against hunger and homelessness.

instructing residents where to head, particularly crucial in more rural areas that had never been popular destinations among visitors. Prominent members of the community travelled door to door with pamphlets promulgating the event: their dependable reputations increased recipients' confidence that the upcoming event was no ruse, while national television and radio broadcasts publicised updates to those willing to trust their provenance. Ironically, many of the informal networks championed by Soviets became conduits of the resistance movement: workers capitalised on the relationships they had built with colleagues as a matter of cultural expectation, and university professors found they could mobilise their students in large numbers, while more broadly, the collectivist mindset helped engender a culture of sharing equipment and assisting others. Though religious leaders had been forced to conduct their affairs underground by the Soviet leadership, they too proved invaluable in sharing plans via their sermons, and their tradition of publishing leaflets helped inspire the pro-independence movements to do likewise.[38] Still, even if every resident of Estonia, Latvia and Lithuania participated, the protest would have minimal impact should international viewers remain unaware of Baltic grievances. And so, visas were arranged hastily for foreign journalists, whose own sudden enthusiasm for visiting a long-neglected part of Europe suggested that they suspected an imminent scoop.[39]

It is easy for us to overlook today how ambitious this plan was at the time. Before one even considers the challenges of motivating participants to set aside any hesitancy about protesting against an authoritarian regime and travel to a potentially distant location for a matter of minutes, the organisers enjoyed little of the technology we take for granted now. Especially in small towns and villages, mobile phones were extremely rare. The World Wide Web was only invented in 1989 and made available to a limited Western public in 1991; although the mathematician and cyberneticist Viktor Glushkov had imagined a Soviet version of the internet called OGAS as early as 1962, his proposal remained severely underfunded until his death two decades later.[40] No internet also meant no smartphones and no social media. In their place, the organisers used walkie-talkies to synchronise their activities, and participants were asked to bring portable radios so that they could receive special broadcasts and listen to political speeches.[41] As many people did not own a car, carpools and free buses were arranged to bring individuals and families from all corners of the region to more remote sections of the route that risked being disconnected. In the event, 2 million people – a good five times more than originally anticipated – ensured that the chain was continuous.[42]

With a start time of 7 p.m., the Baltic Way's architects hoped to include as many Baltic residents as possible. The logic was to capture working adults before they turned in for the evening, and children and teenagers between the end of school and bedtime. Even they were astonished at the buy-in they achieved. Elderly people in poor health, or whose mobility was limited, took pains to attend. Young parents brought their newborns so that they could be a part of this pivotal event. Although some refused, many employers in Latvia and Lithuania allowed their workers to take part of the day off and, in some cases, sponsored company bus rides so that they could participate more easily, while Estonia declared it a national holiday to boost participation.[43] Even if many may have had additional, personal reasons for taking part – from the teenager seeking a pretext to make physical contact with their crush, to the middle-aged parent cautious about participating in their very first act of public rebellion, but assured that they would be surrounded by their loved ones – everybody recognised that this was to be an event like no other, a potentially paradigm-shifting moment in history of which they would be a key component.[44]

For fifteen minutes, enough time for different sections of the chain to be immortalised in photos and film recordings, the Baltic Way's participants held hands, proclaiming their desire for freedom. Facing west so that their backs would be turned to the Soviet Union, the protesters exhibited a shared aversion to Moscow that needed no interpreter.[45] To make sure there wasn't a shred of doubt surrounding their message, many brought national flags and candles tied with Black Ribbon Day's distinctive fabric, and sang national songs that hadn't been heard publicly for fifty years.[46] One Lithuanian pilot, Vytautas Tamošiūnas, dropped flowers from his plane, whose tail he had earlier repainted in his national colours; ever to bear a grudge, the Soviet authorities later revoked his licence.[47] In the Baltic capitals, the leaders of the respective nationalist organisations addressed the demonstrators with rousing speeches tying together the shared fates of the three nations, while a joint statement was issued to foreign media, governments and parliaments demanding that the whole world finally acknowledge their afflictions.[48] Here, they proclaimed, were three nations tied not to a historically contingent Soviet Union, but to a modern and intrinsically free Europe:

> The Hitler–Stalin Pact is still shaping the Europe of today, our once-common Europe. Yet we are convinced that the European community and democratic forces in the East will unite their voices in support of the demand of Estonia, Latvia and Lithuania that the pact together with its secret protocols be denounced and declared null and void from the moment of signing. Only then will Europe divest itself of the last colonies of the Hitler–Stalin era, and the Baltic nations will get the opportunity to determine their own destiny on the basis of free self-determination.[49]

With emphasis placed on the Baltic states' 'independent statehood', no longer could Europe turn its back on three of its countries, annexed with 'overwhelming military force' and 'heavy political terror'. If Europe truly wanted to prove its continued commitment to decolonisation, then, Baltic leaders stressed, it would need to support and recognise the same rights to self-determination in the three forgotten members of the ill-fated League of Nations* as those now enjoyed across Africa and Asia.[50] The Baltics had awakened: had the rest of the continent?

Regardless of these ardent words, it was the chain that helped the Baltic Way stand out from other protests, providing a powerful image that has endured. Through practically connecting the three Baltic capitals' most noteworthy repositories of national identity, and symbolically accentuating the countries' solidarity in the face of decades of oppression, the Baltic Way made Estonian, Latvian and Lithuanian independence not just a conversation topic for Cold War politicians, but a moral issue that could captivate the world. The Baltic states might have been extinguished from the map by authoritarian regimes, causing them to be forgotten by millions across the planet, but they were still here, connected in their traumas, and joined as a single chain.

On the surface, the initial Soviet reaction to the demonstration was as expected. Dismissing the protesters as naive, for 'the fate of the Baltic peoples' was now 'in serious danger', the Central Committee of the Communist Party warned that the 'virus of nationalism' threatened Soviet stability and 'the vital interests of the entire Soviet people'. In response to separatist 'extremism' and 'nationalist hysteria', 'decisive, urgent measures' would likely follow, an implicit threat of military action that spurred some to wonder whether the Baltic states were about to experience a

---

*Established by the Paris Peace Conference following the First World War, this organisation was committed to promoting and maintaining international cooperation and peace. Having joined in 1921, Estonia, Latvia and Lithuania were among the League's sixty-three members, before it was effectively replaced by the United Nations in 1946. By this point, the three Baltic states had been, in the words of the joint statement, removed 'from the memory and map of Europe, from your libraries and textbooks, your sense of justice, your grief and your minds'.

similar communist crackdown to Hungary in 1956 and Czecho-slovakia in 1968, not to mention Beijing just months earlier.[51]

However, other indications were that the Soviet leadership was deeply divided on how best to respond to the Baltic Way. For instance, remarks in the same statement such as 'people should know into what abyss they are being pushed by nationalist leaders' implied that diplomatic rather than coercive means might be used to coax the public into repudiating their commanders.[52] Exemplifying how atomised the Soviets had become across the confederation, the Baltic Way's organisers had actually secured advance permission from local Communist Party authorities to create their human chain, meaning that their actions should never have been a surprise, even if the sluggishness of Soviet bureau-cracy might have delayed any communication between the Baltics and Moscow.[53] (More plausible, though, is that their astonishment owed to the scale and success of the protest, as opposed to its actual occurrence.) Similarly, mere days before the demonstration and hence seemingly in an effort to mollify it, the Soviet Congress of People's Deputies, a new body created by Gorbachev which now constituted the highest organ of Soviet power, finally admit-ted that the Molotov–Ribbentrop Pact did in fact contain secret clauses, but crucially refused to acknowledge the treaty's direct relevance to the Baltic states' annexation.[54] Facing the rising tide of nationalism more directly, Lithuania's Supreme Soviet legisla-ture quickly clarified its own stance: that the Pact's clauses were indeed designed to allow the USSR to occupy the Baltic states illegally, a point confirmed by the frantically backtracking central Congress before the year was out.[55]

In these ways, the Soviet authorities seemed unsure whether or not to appease Baltic nationalists or to curtail their efforts, and if the former, whether it was better to pre-empt or react to their activ-ities. Regardless of what they intended, Baltic activists held firm. Issuing a joint declaration to the secretary-general of the United Nations and a direct plea to US president George H. W. Bush in the weeks after the demonstration, they repeated the key messages of their Baltic Way statement, emphasising that the Molotov–Ribbentrop Pact constituted a flagrant violation of international law that had always been void, and accusing Europe of ignoring the treaty's continued impacts on Baltic citizens.[56] With some encouragement from Bush as well as West German chancellor Helmut Kohl, the Soviet Union soon softened its stance.[57] For a moment, it seemed as if perestroika and glasnost had finally supplanted the Stalinist purges of the past.

Soon, however, the Soviet Union reverted to type; for whereas the Baltic Way was crucial to Estonia, Latvia and Lithuania's interconnectivity, for the Eastern Bloc as a whole it was a major catalyst of instability. Just as Poland's Solidarność (Solidarity) movement had animated anti-communist movements in Central and Eastern Europe nine years earlier, the demonstration proved that millions in the Baltic states craved independence and were now willing to challenge Soviet hegemony in order to realise this objective. Though they only represented a tiny proportion of the USSR's population, these protesters risked initiating a domino effect and hence posed an existential threat to the union as a whole. Indeed, ensuing events demonstrated that this chain of people was able to create the figurative cracks that would expedite the confederation's eventual collapse (as well as indirectly producing more literal cracks in Berlin).

Having already declared their sovereignties and the primacy of their national laws, all three Baltic states held free democratic elections for the first time in February 1990. In none of the three elections was the result close, with pro-independence candidates winning comfortably.[58] On 11 March 1990, Lithuania became the first Soviet state to declare its independence, a proclamation Gorbachev predictably, but unsuccessfully, tried to overrule.[59] Under two months later, Latvia followed, while Estonia proceeded with greater caution at first, opting to abandon its Soviet Socialist appellation. Refusing to take matters lying down, the USSR responded with intimidation – and not simply by mobilising its troops. Recognising its mastery over Lithuania's lifelines, in the first instance the Soviet Union imposed an economic and energy blockade which would cut off the small nation's supplies of oil and gas, obstruct its financial transactions and restrict its ability to import raw materials and goods for two and a half months.[60] Later it employed more physically coercive tactics, as peaceful protesters in Lithuania and Latvia experienced first-hand in January 1991. Following the Lithuanian Supreme Council's refusal to reinstate the Soviet constitution with immediate effect, an ultimatum was issued straight from Gorbachev's office in Moscow. Soviet troops with KGB support began to seize Vilnius's key political and media structures, including the National Defence Department building, the Press House and the TV Tower, still the country's tallest structure. Determined to defend their state institutions, thousands of Lithuanians flocked to these buildings and formed human chains of a new kind, barricades of people that held firm even as the army ploughed its tanks into some protesters and shot bullets at others. Over the course of three days, more than a dozen unarmed civilians lost their lives, while more than a thousand were wounded.[61]

Reflecting their newfound solidarity with their southern neighbours, Latvians responded to the onslaught in Vilnius as well as the threats they too were receiving from the Soviet leadership by erecting barricades of their own in Rīga.[62] Having already seen their printing house occupied by Soviet special forces barely two weeks earlier, which now churned out a steady supply of pro-Kremlin propaganda, Latvian journalists and politicians called on their compatriots to anticipate the arrival of Soviet tanks by blocking access to all nationally important sites.[63] People from across Latvia arrived in the capital with material and machinery, laying logs, construction debris, concrete blocks and other obstacles around the Supreme Council, the Council of Ministers and television and radio buildings, as well as at key junctures of connective infrastructure such as main roads and bridges.[64] Fishing ropes were stretched across the Daugava River to tangle Soviet ship propellers.[65] To keep spirits up, musicians gave live performances while bonfires not just offered warmth from the chilly winter air but, in evoking traditional midsummer celebrations, acted as a symbol of Latvians' resolve, too.[66] As with the relatively media-savvy organisational techniques behind the Baltic Way, radio was used to coordinate demonstrations, meetings and eating and sleeping arrangements, while the press was tasked with keeping the subsequent Soviet offences in the national and international mediascape.[67] In fact, the Soviets' immediate use of automatic weapons in Rīga and Vilnius against people who had been instructed by their leaders to arrive unarmed was so shameful that even in Moscow, more than 100,000 Russians marched to the Kremlin calling for Gorbachev's resignation.[68] The Latvian victims of this two-week period continue to be remembered in the form of bonfires and public gatherings on 20 January every year, while more than 30,000 people have been awarded a commemorative medal for their bravery in defending Latvia's independence.[69] Riga is also home to a small museum and memorial dedicated to this critical event, and in 2023 it dedicated the entirety of January to barricades remembrance.[70]

With the world's eyes belatedly but quite firmly cast on the Baltic, where Soviet forces continued to attack new border posts while Gorbachev spuriously denied personal responsibility for his forces' brutality in Lithuania and Latvia, the USSR's international reputation was well and truly shot.[71] Seizing the opportunity, Lithuania held a new referendum on 9 February. Noticing the public's overwhelming support for independence, just two days later, Iceland became the first country to formally recognise Lithuania as a sovereign independent state,* prompting the Soviet Union to

---

* Moldova had actually recognised Lithuania's independence as early as May 1990, but its sway was appreciably limited by the fact that it, too, was still

recall its ambassador to this Nordic nation.[72] With Latvian and Estonian independence referenda subsequently yielding similar (and given these nations' larger ethnic Russian populations, somewhat surprising) results, Gorbachev scrambled to hold a vote of his own, asking citizens throughout the USSR whether they wanted the union to be preserved as a renewed federation of sovereign republics.[73] The three Baltic states as well as Armenia, Georgia and Moldova* duly boycotted what proved to be the confederation's one and only national referendum in its history, galvanising their aspirations for complete independence, even if Gorbachev could point to the remaining republics' widespread approval of his reformation plan.[74]

Still, a new challenge in August 1991 ultimately saw the end of Gorbachev's premiership: a coup in the capital by hardline communists, a collection of military leaders and KGB agents who objected to the country's reforms and political decentralisation.[75] (Less known elsewhere, for most eyes were on Moscow, is that in a reprise of the 'January Events', Soviet troops simultaneously started seizing dozens of important buildings in Vilnius, which prompted Lithuanian civilians once again to form a human shield around their parliament building.)[76] Though the poorly organised revolt in Moscow was unsuccessful in installing a new regime, Gorbachev's authority would never recover.[77] The putative great reformer and the previous year's Nobel Peace Prize winner resigned as the Communist Party's general secretary (though he remained Soviet president until late December) a day after the second anniversary of the human chain.[78] His successor, Boris Yeltsin, immediately acknowledged the Baltic states' independence, and called on others to do the same.[79] Within two weeks,

---

a part of the Soviet Union at this time. Incidentally, later that year, Iceland initiated international recognition of Latvia and Estonia as well.

* Minus Abkhazia and South Ossetia in Georgia, and Transnistria and Gagauzia in Moldova, all four of whose leaderships remain pro-Russian.

all three were officially Soviet republics no more; within a further two, all three had been admitted into the United Nations.[80]

Dissolved before the year was out, the USSR was replaced in effect by a new Commonwealth of Independent States (CIS) comprising those new states that remained interested in cooperating as part of a Russia-centric supranational organisation. In a sign of their westward orientation and contempt for the Soviet Union, Estonia, Latvia and Lithuania are the only former Soviet states never to have joined,* and since 2004 they are also the only such countries to have been admitted to NATO and the European Union. All three have demonstrated remarkable progress in guaranteeing the types of individual freedoms inconceivable in the USSR: the non-profit organisation Freedom House ranks Estonia close to Switzerland, Japan and Iceland, and ahead of Germany and the United Kingdom, while Lithuania and Latvia rate similarly to France. Russia, meanwhile, sits below the Republic of the Congo, Venezuela and Chad.[81] Additionally, the Baltic states are today the wealthiest post-Soviet nations and boast thriving business sectors: the World Bank ranks Lithuania as the eleventh easiest country in the world to conduct business, while Estonia and Latvia, but not Russia, also make the top twenty.[82] However, the post-socialist transition has also been anticlimactic for many residents, who bemoan capitalism's tendency to produce 'losers' as well as 'winners', a phenomenon felt most acutely during the 2008–9 financial crisis.[83] In particular, many ethnic Russians complain that they now represent a discriminated-against class, being far more likely to live in poverty than ethnic Latvians and Estonians, as well as often facing significant dilemmas over their right to continue living in the Baltic, for Latvia now uses proficiency in the

---

* Ukraine and Turkmenistan did agree to the formation of the CIS, but never ratified its charter, making them associate rather than full members. Since their respective invasions by Russia, Ukraine and Georgia have terminated their involvement in the CIS.

Latvian language (which many ethnic Russians do not speak) as part of its test for permanent residency, while Estonia, generally, does not permit dual citizenship.[84]

Such tensions within the Baltics are replicated at the international scale, for the three states' relationships with Russia have remained icy ever since the Cold War. Having inherited the USSR's antipathy towards NATO, an alliance originally created in part to contain the Soviet Union and hinder its expansionist ambitions, Russia's paranoid suspicion of the trio has only intensified since their accession to this organisation. And although the USSR eventually recognised the illegality of the Molotov–Ribbentrop Pact, much of Russia's current leadership, including President Vladimir Putin, still views the treaty as legitimate, providing a potential pretext for a future invasion.[85] Watching Russia brutalise Ukraine, another post-Soviet state that has been working hard to repudiate its historic occupier, it is certainly understandable that many in the Baltics, pro-Western countries sitting right on Russia's doorstep, fear that they will be next. (As both friend and extension of their own security, the Baltic states have accordingly offered Ukraine more aid relative to GDP than any other country bar Denmark.)[86] After all, they know Russia's tendencies better than anyone.

Even if Russia is conscious that the Baltic states will never choose to return to its sphere of influence, it recognises that by sowing discord internally, it can fray their links to the West. For this reason, the three states remain highly vigilant about Russian meddling in their political systems, cultural institutions and media and, with an eye to the entire region's future stability, have proved proactive (and brave) in challenging their much bigger neighbour whenever it and its chief partner Belarus overstep their bounds.[87] Latvia embodied these traits in the wake of Russia's escalation of hostilities in Ukraine in 2022, first nationalising Rīga's Moscow House – a cultural centre for Russian speakers which, with financing from the Russian government, was trying to justify Putin's actions in Eastern Europe – and then sending the money raised from selling the property to Ukraine to support its war effort.[88] Looking north and west, the Baltic trio have welcomed Finland and Sweden's recent accession to NATO, a move that has created a contiguous line of member states from the Arctic to the Mediterranean and theoretically strengthened their combined ability to resist Russian threats in the Baltic Sea. That said, the presence of the tiny but militarily mighty Kaliningrad exclave continues to be just one source of consternation in a body of water where internationally connective critical infrastructure, including gas pipelines and communications cables, is constantly vulnerable to sabotage.[89]

United in their experiences of occupation and liberation, and conscious of the power of the media in broadcasting their concerns, ultimately Estonia, Latvia and Lithuania are aware that if they (as well as regional allies such as Finland) are to survive as sovereign states, they must continue to draw attention to dubious actions in and around the Baltic Sea whenever they arise. Though they adjoin Russia geographically, perceptually the Baltic states could scarcely be more distant. Yet just as this part of the world epitomises how charged borders can be, it also exemplifies how the very same entities don't have to imply emotional disconnection. For five decades, the Baltic nations' borders were concealed within the Soviet behemoth, and their people's sense of identity was browbeaten until it seemed at risk of disappearing. However, through their conjoint protests, Estonians, Latvians and Lithuanians transformed the geography of their region, acknowledging the sanctity of the boundaries officially separating them without allowing these to undermine or preclude their collaboration, while simultaneously entrenching the divide they perceived from their occupier to the east.[90]

Such a logic persists to this day. Aided administratively by the Schengen Area's principle of free movement, the borders between the Baltic states welcome enduring cooperation (and additionally act as symbols of their shared independence struggle), at the same time as their eastern boundaries operate as formidable barriers to movement from Russia. In this reimagined geography, it is more than a little telling that the three-nation chain the Soviet Union had annexed as a geostrategic buffer now acts as a first line of defence for the countries to the west. Once historically discrete, these nations know that they are stronger together than apart: in one of the planet's most intense geopolitical hotspots, their success in enhancing their connections – both internally and with the West – is critical to Europe's future.

More viscerally than perhaps any other event in history, the Baltic Way gave meaning to the common adage 'power of the people' and proved that the simple action of holding hands has the potential to overcome traditional national boundaries and change the world. While many protests are dismissed by their opponents as somehow fanatical, here was a fundamentally inclusive form of patriotism, concerned less with whether an individual identified as Estonian, Latvian or Lithuanian than that they shared a common goal of freedom from despotism. Though the Kremlin briefly tried to delegitimise the Baltic Way by portraying its organisers as fascists and bourgeois nationalists, its charges were patently inconsistent with the humble and heart-warming act of standing hand in hand.

Of course, the human chain owed much of its impact to its emotional resonance, as people often with little in common on a personal level set aside any differences to convene and interlock themselves with one another, creating a united entity of which they were all an essential part. Looking left and right, what they saw was non-violent solidarity, a visual expression of the fact that together, they were one. But no less compellingly, the human chain offered unique *practical* advantages as a form of protest, essential in confronting a far bigger power with a reputation for intolerance. First, such a demonstration is characterised by flexibility, regardless of the meticulous planning involved in the case of the Baltic Way. An opponent would find it impossible to disperse protesters distributed over hundreds of kilometres, for though single sections of a line might be broken, the majority can remain intact.[91] Second, given that the individuals involved must travel to and from disparate places, they are able to converge spontaneously in the eyes of the authorities, precluding any efforts to obstruct them. Third, a human chain is a physical entity whose visually arresting shape demands to be seen, reported and discussed, in a way that petitions, for instance, lack. Significant in this respect is that even though petitions were signed denouncing the Soviet occupation, they provoked much less of a reaction from the Muscovite leadership and hence were simply the accompaniment to the chain. And fourth, a human chain is inherently unthreatening. Opponents would be hard pushed to justify a physical response to people who are both stationary (unlike a march) and holding hands (and therefore probably not holding weapons). This advantage was later recognised by the Lithuanian leaders of the 'January Events', as they called on the public to hold hands and lock rather than carry arms in their attempts to protect Vilnius's most important buildings, according to the principle that with connection comes protection. Whether or not they ever articulated them, the Baltic Way's organisers clearly understood that human chains have traits uniquely suited to protest, and in this respect, their concept was ingenious.

Certainly, the advantages of this distinctly connective form of political demonstration have since been embraced by subsequent protesters seeking to resist forces far larger than themselves. In 2004, a human chain extended across the length of Taiwan to commemorate an anti-government uprising that was violently quashed in 1947 and to oppose the People's Republic of China's continued military threats towards this disputed territory.[92] In 2019, China again became the target of such dissent, this time in Hong Kong, where pro-democracy protesters sought to raise awareness of a proposed law to allow criminal suspects to be extradited and tried on the mainland.[93] And in the interim, Catalan demonstrators in 2013 created a 400-km (250-mile) *Via Catalana* ('Catalan Way') to demand their autonomous community's independence from Spain.[94] Though these events have had varying levels of success so far in realising their objectives, in each case their organisers recognised the important legacy provided by three Baltic countries whose collective population is less than 1 per cent of Europe's total. If these small nations could stand up both figuratively and literally to their oppressor, create a new, liberated geography and alter the course of history both within and beyond their boundaries, then why not us as well?

Estonia, Latvia and Lithuania each continue to write their experiences of collaborative rebellion onto the landscape, ensuring that the event that returned them to the map is never far from citizens' minds. Today, museums, memorials and events commemorating the Baltic Way are scattered across the route, evincing how the chain concerned obscure rural locations and not only larger towns. Still, the simplest yet the most impactful expression of this region's emotionally connected geography can be found in the three capitals, each of which boasts one of three matching footprint tiles, signifying one of the locations where protesters once stood. Even as remembrance of the Baltic Way continues to evolve, its carefully drawn route remains fundamental to how people engage with this unusually corporeal and internationally connective form of peaceful protest. On its twentieth anniversary in 2009, thousands of runners took part in a relay starting from either Tallinn or Vilnius and meeting to celebrate together at the Freedom Monument in Rīga.[95] Ten years later, a rally of vintage cars, some of which had been used to shuttle participants for the original chain, retraced the route from south to north, while concerts, bike rides and photo exhibitions drew in individuals from across the region.[96] However, politics is never unanimous. As the Baltic Way is so central to all three countries' national narratives, attempts to repurpose it for unpopular causes can attract widespread condemnation, as exemplified by segments of the 2021 chain seeking to protest a different set of restrictions: those associated with the COVID-19 pandemic.[97]

The conceptual power of the Baltic Way has not faded, especially in Eastern Europe. The fate of Belarus, Latvia and Lithuania's south-eastern neighbour and a country whose pro-Russia president Alexander Lukashenko is often unflatteringly described as 'Europe's last dictator', is today a particular topic of interest to many in the Baltics, especially given that its authoritarianism and human rights violations (plus its support for Russia's invasion of Ukraine) offer an uncomfortable reminder of recent Baltic history. For this reason, following Belarus's sham presidential election in 2020, which was won as always by Lukashenko, not only did Belarusians bravely take to the streets to form human chains in protest, but people in all three Baltic states also momentarily disregarded their COVID-related fears to demonstrate in the now traditional manner on 23 August once again. Most notably, 50,000 people in Lithuania, waving the red-and-white flag popular among Lukashenko's pro-democratic opponents, held hands between Vilnius and the village of Medininkai, extending the old chain all the way to the Belarusian border and thereby bestowing upon Belarus the title of honorary Baltic state.[98] Far from being consigned to history as an interesting but now immaterial moment from a past epoch, the Baltic Way was thus alive once again, a 'Freedom Way' for those still fighting totalitarianism.

In turning a regionally significant but sombre day of mourning and loss into a birthday for a new post-Soviet world, the Baltic Way embodied the power of geographical connection as a vehicle of resistance. After all, at the heart of any kind of resistance is the belief that circumstances *can* be altered. Through our actions, connecting seemingly disparate places, we have the capacity to take control over the future of our planet, and repair even its severest ruptures. Inscribed in UNESCO's Memory of the

World Register in 2009, the event continues to symbolise values of peace and solidarity, and exemplifies the capacity of non-violent protest to overcome prejudice and suppression. Both palpably and perceptually, the protest demonstrated how geography is never inexorable, and provided a timely reminder that borders, the official divisions between states, are unable to tell the whole story. Just as a specifically Baltic history and identity now tie together three nations that, prior to the Soviet Union, existed quite separately from one another, so too might we find affinities elsewhere, rooted in common experiences and shared connections to a land that, despite appearances, is by no means immutable.

6

Restoration: The Great Green Wall

The Great Green Wall is a new world wonder in the making ... It shows that if we work with nature, rather than against it, we can build a more sustainable and equitable future.

Amina Mohammed, UN deputy secretary-general[1]

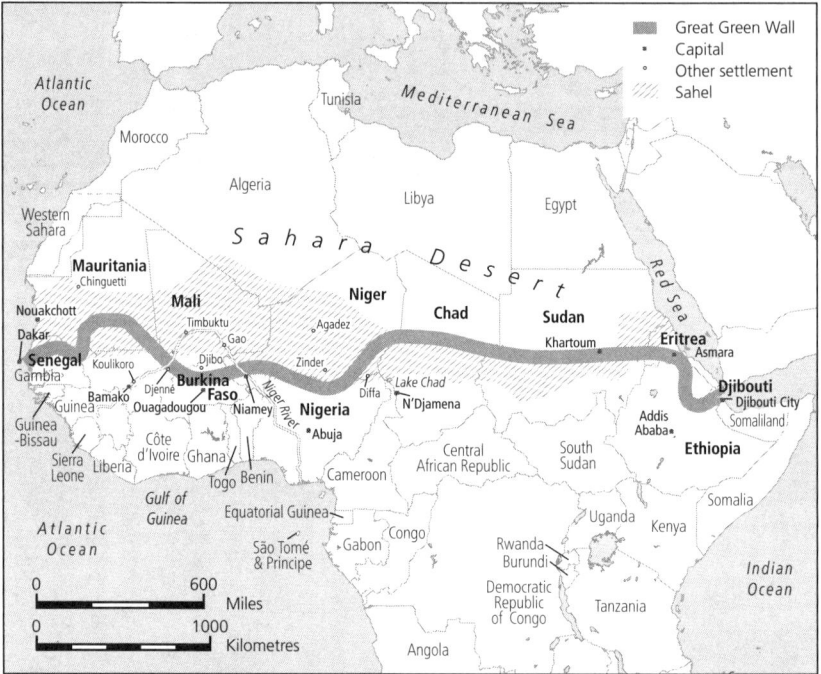

Great Green Wall
Capital
Other settlement
Sahel

Atlantic
Ocean

Mediterranean Sea

Tunisia

Morocco

Algeria

Libya

Egypt

Western
Sahara

S a h a r a   D e s e r t

Red Sea

Mauritania

Chinguetti

Mali

Niger

Chad

Sudan

Eritrea

Nouakchott

Timbuktu

Gao

Agadez

Khartoum

Asmara

Dakar

Djibo

Zinder

Senegal

Koulikoro

Djenné

Burkina
Faso

Niger River

Diffa

N'Djamena

Djibouti

Djibouti City

Gambia

Bamako

Ouagadougou

Niamey

Nigeria

Lake Chad

Somaliland

Guinea

Guinea
-Bissau

Côte
d'Ivoire

Ghana

Abuja

Addis
Ababa

Ethiopia

Sierra
Leone

Liberia

Togo

Benin

Cameroon

Central
African Republic

South
Sudan

Somalia

Gulf of
Guinea

Equatorial Guinea

Congo

Uganda

Kenya

Indian
Ocean

Atlantic
Ocean

São Tomé
& Príncipe

Gabon

Rwanda
Burundi

Democratic
Republic
of Congo

Tanzania

0   600

Miles

0   1000

Kilometres

Angola

The heat is oppressive. Weary birds drift across the sky, yearning to escape the blistering sun. Below, the soil is being baked to diamond hardness, its value diminishing with every moment. Grasses dry to matchsticks; skinny goats teeter and swoon. Hunger and despair are etched on the haggard faces of people who, with the shifting sands of time, reap little more than an early glimpse into an apocalyptic future, as a rapidly changing climate drives new population movements, new pressures and new conflicts.

Stretching for approximately 6,000 km (3,750 miles) from the Atlantic Ocean to the Red Sea, the Sahel is a continuous belt of semi-arid land between the dry Sahara to the north and the wetter savannahs to the south.[2] Depending on one's definition, the Sahel intersects a good ten countries, including parts of Senegal, Mauritania, Mali, Burkina Faso, Niger, Nigeria, Chad, Sudan and Eritrea, and arguably The Gambia, Guinea, Guinea-Bissau, Cameroon, Ethiopia and Djibouti as well. Clues of an extraordinary past remain in the architecture of various medieval cities along the region's main trans-Saharan trading routes, including the mosques of Agadez in modern-day Niger and Gao and Djenné in Mali, whose adobe exteriors adorned with protruding sticks of rodier palm cause them to resemble eclectic hybrids of cacti and French pastries. For the majority of the Sahel's inhabitants today, however, the fabled histories of Chinguetti in Mauritania (a popular gathering place for Muslim pilgrims heading to Mecca, and an important repository of Quranic texts) and Timbuktu in Mali (a flourishing and profoundly interconnected hub for scholars and traders that far outstrips its mythologised 'far-off' reputation among Europeans) are of little consequence. Marred by drought and poverty, disease and bloodshed, this once-illustrious strip of land has metamorphosed into the world's biggest crisis zone.

For as much as the Sahel refers geographically to the various lands joined together by this 'shore' at the desert's edge, its contemporary connective relevance is far more fluid. The region ties together more than 100 million people who, by virtue of their very existence, face challenges unimaginable almost anywhere else on the planet.[3] Although the popular but overly simplistic notion of a creeping desert obscures the specific geography of land degradation, at the heart of the matter is desertification: the Sahara is now up to 18 per cent larger than it was a century ago.[4] It is easy to blame this rapidly changing geography on climate change. Temperatures in the Sahel are rising one and a half times faster than the global average and are projected to increase by another 3–5°C by 2050.[5] Considering that in some places temperatures can already push 50°C, this region is quite literally feeling the heat more than anywhere else on Earth. Coupled with an abbreviated summer rainy season, especially in the west, and a winter dry season which is seeing increasingly intense harmattan winds from the Sahara bring thick clouds of dust, forests and fields are degenerating before people's eyes.[6]

However, climate change is not the only impetus behind the growing desert. Alongside the Sahel's multi-decadal increase in temperature and decline in rainfall, human activities have left once-vegetated ground bare. Whether directly or not, an underlying cause is the Sahel's rapid population growth, as dazzling as the sun overhead. By virtue of a well-founded fear that many children won't survive until adulthood, a widespread belief that more children means more chances of economic success, and religious and cultural expectations that champion large families over family planning, this is the most fertile region in the world – in terms of demographics.[7] At a little under seven children per woman, Niger's fertility rate is three times above the global average and over 70 per cent higher than the average of the world's *least* developed countries.[8] Add to that the fact that nearly two-thirds of people in the Sahel are under twenty-five, most of whom still have plenty of child-rearing years ahead of them, and it's clear that the region faces near-irresolvable dilemmas, both demographic and environmental.[9] Continued population growth here is a given: according to current projections, the region could be home to 330 million people by 2050, and double that by 2100.[10] The problem at hand is that productive land isn't expanding at the same rate. Worse, due to a combination of climate change and weak government regulations and oversight, it's diminishing. And so, with increasing pressure placed on fewer and fewer shreds of viable terrain to feed more and more people, deforestation, over-cultivation and overgrazing now run rampant, driving the region's natural resources to breaking point and exposing the nutrient-rich topsoil to the elements.[11] When the rain finally falls – which it tends to do as torrential downpours – it can then prove less a comfort than a curse, washing away the essential topsoil, engulf-ing exhausted communities in floodwater and attracting biblical swarms of desert locusts so dense they are capable of forcing pas-senger planes to divert.[12]

Confronted with this taxing alternation between long droughts and severe deluges, intensified by voracious insects and population pressure, many have opted to pursue a real-life twist on the Exodus story by migrating to just a few relatively arable provinces within the Sahel – or to urban or coastal areas beyond.[13] However, rather than finding their Promised Land, most people continue to face a series of modern plagues, in a broad region that since the 1980s has assumed the unwanted but tragically fitting nickname the 'Hunger Belt'. Whereas for centuries herders tended to be welcomed by southern farmers because their animals would fertilise their fields, today violent clashes over increasingly scarce resources have become all too frequent, particularly where legal rights to land are inconsistent or unclear.[14] Catalysed by the rise of ethnically based militias,* Islamist groups including Boko Haram and Islamic State have expanded across the Sahel, seeking to exploit these fragile states' general lack of political stability.

While it's true that people rarely join these groups for doctrinal or ideological reasons – more commonly they are youths desperate for income, status, belonging and protection – the repercussions are nevertheless severe.[15] In destroying schools, places of worship, markets, homes and more in scorched-earth offensives, not only have these organisations exacerbated the Sahel's development challenges directly, but they have also provoked increasingly aggressive counter-terrorist responses prone to driving even more people to jihadism – most notoriously in Burkina Faso.[16]

Besides connecting millions of the Sahel's residents through their shared experiences of adversity and anguish, this medley of desertification, destruction and disorder has undermined the prospect of meaningful gains being made in other key areas of society – the sort of advances that would help alleviate some of the problems at the region's root. Unrelenting violence has both exacerbated and been fuelled by local communities' difficulties in accessing essential resources around the redundantly named Lake Chad ('Chad' literally translates as 'lake'), a water body that shrank by more than 90 per cent between the 1960s and 1990s due, primarily, to climate change.[17] In Burkina Faso, water and healthcare services have become entangled in combatants' military strategy: each assault ensures that these necessities remain outside the reach of a growing number of people.[18] A pervasive

---

* Two prominent examples are the National Movement for the Liberation of Azawad (MNLA), a movement of mostly nomadic Tuareg people who in 2012 felt sufficiently marginalised by the Malian state to fight for an independent territory, and the predominantly Muslim Fulani herders, feared by many Christian communities in northern Nigeria for their periodic attacks on villages, churches and farmers.

combination of poverty and insecurity has meant that the Sahel remains particularly vulnerable to various diseases that are now relatively rare elsewhere, including malaria, yellow fever, cholera and meningitis, while the risk of encountering assailants has prompted many pastoralists to alter their transhumance routes, affecting the balance of fragile ecosystems. A number of native mammals have become critically endangered (for instance, the addax antelope and the dama gazelle) or even extinct (the bubal hartebeest) through a combination of over-hunting, competition with livestock and armed conflict. Humanitarian aid workers have been periodically targeted by armed groups in regions such as Diffa in south-eastern Niger, preventing those in need from receiving vital support.[19] Tragically, tens of thousands of people have been murdered across the Sahel in recent years, while millions more, desperately searching for physical safety, and water and food security, have been forcibly displaced both within and across its various national borders.[20]

Epitomising the myriad challenges facing the Sahel, it is telling that with the exception of Nigeria, which is only grazed by the region in its far north, every single state in this contiguous belt is regarded by the United Nations as being among the world's least developed countries.[21] Yet beyond the region's trifecta of interconnected challenges – climate change, overpopulation and conflict – a less widely recognised factor in shaping its past, present and future is the role of connection of a more specifically geographical kind. For centuries, nomadic and semi-nomadic groups have journeyed vast distances across the Sahel and Sahara, seeking to maximise the seasonal pastures in different regions without over-exploiting them.[22] Merchants and their camel caravans have traditionally relied on Indigenous groups' familiarity with the desert, following meticulously defined itineraries between oases to traverse some of the planet's most arid natural landscapes.[23] Mansa Musa, a fourteenth-century Malian emperor whose control of the trans-Saharan gold and salt trades enabled him to become wealthier than perhaps any other person in history, remains legendary for his lavish gift-giving along popular trading and pilgrimage routes from modern-day Guinea to Mecca. A few of the mosques constructed or upgraded throughout the Sahel under his patronage are still in use today.

While these relatively small numbers of periodic travellers have historically connected the Sahel in a somewhat tenuous way, today an ambitious initiative seeks to materialise the region's long-standing ties and bring overdue advantages to millions. In contrast to the many walls, both past and present, from Berlin to Babylon, whose purpose is to divide and exclude, this project, the Great Green Wall, shows that these structures can have a productive, future-facing function instead. Both a barrier against desertification and a unifier of ruptured societies, the enterprise encapsulates the latent potential of geographical connection in restoring and reviving our world. Just as Musa's exceptional wealth and diplomatic ties seven centuries ago should remind us that the Sahel's contemporary issues need not be inevitable nor permanent, the Great Green Wall demonstrates that, through mindful action, commitment and collaboration, we can transform our broken surroundings. Immense and inspiring, idealistic yet prudent, a vision of literal connectivity cutting across one of the world's most fractured regions, this international initiative captures the imagination like few others.

Meandering from west to east across sub-Saharan Africa, the Great Green Wall is an ongoing project to create the world's largest living structure, a green buffer over three times longer than the world's current record holder, the Great Barrier Reef. At this endeavour's core is the earnest aim of restoring 100 million hectares of the planet's most degraded terrain by 2030, along an uninterrupted, 8,000-km (nearly 5,000-mile) course connecting Senegal, Mauritania, Mali, Burkina Faso, Niger, Nigeria, Chad, Sudan, Ethiopia, Eritrea and Djibouti.[24]

To the potential disappointment of anyone expecting one day to see a colossal barricade of trees standing stoutly at the edge of a vast desert, the project is actually far more nuanced. Although the original plan was indeed to connect Africa's west and east coasts with a continuous 15-km (9-mile)-wide wall of drought-tolerant trees, the Great Green Wall has since evolved into a linear mosaic of productive landscapes, including forests as well as grasslands, savannahs, farms and community gardens. Still, don't think that this looser interpretation of a green wall detracts from its connective power. Even though the new vision may sound less grandiose, not only are its essential values and fundamental objectives to restore the Sahel and improve local people's livelihoods very similar, but its construction is far more realistic, too. Practicalities are paramount: a wall broken up by long treeless gaps isn't really a wall.

In order for this project to succeed, each and every section needs to be intact. The Sahel, after all, is a necessarily interconnected region, where land degradation locally can quickly set in motion a domino effect of migration, resource pressure, violence and extremism across a much wider area, damaging additional lands in the process. By contrast, regeneration promises to spark further benefits, rendering it the first step to tackling the Sahel's grievous problems. The cultivation of food ought to act as a safeguard against malnutrition, still a life-threatening issue for millions of people in the Sahel.[25] With enhanced food security, as well as the intended creation of 10 million green jobs in some of the world's most destitute rural areas, regional security will hopefully follow: improved living conditions will provide less sustenance for the growth and spread of extremist groups, whose recruitment largely depends on exploiting citizens' vulnerability, and whose reach is enabled by the porousness of many of the Sahelian nations' borders. And as if these remarkable advantages weren't enough on their own, the Great Green Wall's relevance stretches far beyond the Sahel. Through targeting the capture and storage of 250 million tons of carbon, the region – which happens to contribute the least to global carbon emissions relative to its population size, yet is one of the most sensitive to climate change's impacts – would in fact be doing the whole world an undue favour by mitigating the planet's greatest existential threat.[26]

Given these varied and intrinsically international benefits, it is unsurprising that the Great Green Wall has captured the interest of a wide range of partners in the Sahel and beyond, including the United Nations, the World Bank, the African Development Bank, the European Union, the International Union for Conservation of Nature, foreign governments, non-profits and private actors. It is the butterfly effect manifested as a multi-billion-dollar initiative, uniting a miscellany of groups who recognise that one small modification to a local landscape – for better or for worse – can prompt all manner of impacts thousands of kilometres away. In the Sahel, everything is connected – and those of us elsewhere in the world aren't entirely detached from the region, either.

With all eyes suddenly on the Sahel, the question then becomes, who leads this revolutionary initiative? The answer, so obvious on the surface and yet sadly so radical in modern history, is Africans. For decades, sub-Saharan landscapes tended to be managed according to the priorities and assumptions of powerful people in distant places,* uprooting Sahelian communities' distinctive management of familiar lands in the process. Most significantly, the western Sahel's colonial period (which saw the now-independent nations of Senegal, Mauritania, Mali and Burkina Faso, together with Guinea, Côte d'Ivoire and Benin, incorporated into the authoritarian federation of French West Africa from 1895) saw a hitherto regionally oriented commercial network, which tied together a large number of historic Saharan and Sahelian settlements such as Zinder, Djenné and Timbuktu,

---

* An important exception was Ethiopia in the east, which, apart from a brief period of Italian occupation from 1936 to 1941, was never colonised.

replaced by a more hierarchical model of connectivity designed to entertain foreign interests. Immediate access to potential water routes suddenly became a key determinant of which places would grow and which would shrink. Dakar on the Atlantic coast and Bamako and Niamey on the Niger River, until then all relatively minor settlements, henceforth enjoyed particular privileges as new centres for international trade and military power.[27] It is no coincidence that after independence in 1960, each was chosen as a national capital (by Senegal, Mali and Niger, respectively). To help consummate (and, via the police and military, supervise) this modern connective system, the French also, with some difficulty, built a railway specifically to bring goods all the way from their Niger River terminus at Koulikoro, via Bamako to their new colonial capital, Dakar.[28] Though this railway's primary freight – peanuts – might seem far humbler than the minerals prioritised by Europeans in Mozambique and its neighbours, these legumes' mass cultivation would soon have a practically anaphylactic impact on Sahelian landscapes.[29]

Certainly, in order to introduce monoculture cash crops demanded abroad, including peanuts as well as cotton, lands that had been communally owned under a customary authority had to be modified and were eventually decimated.[30] Having cleared forests en masse, colonial officials allowed the land to become desiccated, devoid of topsoil and resistant to rain.[31] France's new forestry laws proved similarly counterintuitive: they made all trees state property and prohibited colonised communities from cutting them down or even pruning them, so trees enjoyed less, rather than more, protection. Any effort to manage and care for them risked time in jail.[32]

In this fashion, most of the Sahel was transformed into a mere supplier for European markets – and seldom a robust one at that. Yet despite the patent unsuitability of French land-use procedures to the Sahel's hot, arid climate, colonial laws and ideologies proved influential even after independence. Successive national governments chose to replace forests with farms to increase food production, and ensured that trees and crops were kept separate.[33] Though well-meaning, international interventions also tended to prove unpopular and ineffective, largely because of the doctrinaire and rigid way in which these generally capital-intensive and machinery-driven methods were imposed on local communities without consultation.[34] Notably during the 1960s–70s in northern Burkina Faso's Yatenga Province, the earthen terraces constructed over entire catchments were soon either ignored or destroyed by farmers resentful that these bunds prevented water run-off from reaching their fields.[35]

Only in the 1970s–80s, a period that saw increasingly frequent, long and severe droughts rooted in a calamitous combination of overpopulation, overgrazing, deforestation, monocropping, crop failures and climate change, did alternative practices begin to be taken somewhat seriously and, with the support of empathetic non-government organisations, circulate beyond their local contexts.[36] The willingness of Oxfam (a British charity established to feed the starving citizens of Greece during the Second World War) to listen to farmers in Yatenga and shift its focus from tree planting to food production via the construction of more accurately aligned stone bunds embodied the onset of a new era of international collaboration.[37] From this point on, Sahelian communities would belatedly enjoy the primary say over their lands, while their foreign partners would be responsible for maximising their chosen practices' efficiency and helping these become more widespread.[38]

The Great Green Wall can be regarded as the culmination of this shift in emphasis, from the foisting of land-use practices typical in France, to timely interventions nevertheless conceived according to an outsider's assumption of Sahelian needs, to, finally, a project designed for Africans by Africans. That said, as challenging as it is to transform lands, it can be just as difficult to change mindsets. The Great Green Wall epitomises this unfortunate reality, because in its original, tangibly connective form, the project actually better resembled a distinctly foreign and flawed approach to restoring arid landscapes: mass tree-planting.

The belief that trees have an inherent ability to cure all sorts of environmental problems has prevailed for two and a half centuries in Europe, and eventually came to inform environmental thinking far more widely. The Great Green Wall's planners might have looked north of the Sahara, where since the early 1970s Algeria has been struggling to grow a durable *barrage vert* (green dam) in line with a similar French proposal from well over a century earlier. Alternatively, they might have learned from the central United States, where a plan to grow a 'great wall of trees' from Texas to the Canadian border in the 1930s left behind some locally significant windbreaks, but not the continuous, permanent shelter belt envisioned.[39] China should also have provided a warning: in its northern regions long strips of trees were planted from the 1950s to reduce the intensity of asphyxiating dust storms, stabilise moving sand dunes and, with exemplary totalitarian flair, 'prove' that nature can be overcome – but China has mostly found this goal more elusive than the Yeren, China's analogue to North America's Sasquatch.[40]

Considering this quite expansive geography of trying to develop tree barriers against the desert, it is little wonder that the Sahel has also long been regarded as a potential realm of re-greening. In the first instance, British environmentalist Richard St Barbe Baker proposed a 50-km (30-mile)-deep band of trees as early as 1952, and although his 'green front' vision was never realised, various non-profit organisations have since aimed to reforest other parts of the African continent, in many places with great success. A little more than three decades later, Burkina Faso's Marxist president Thomas Sankara became the first African leader publicly to champion cross-border re-greening projects in response to the Sahel's desertification, while chiding the leaders of powerful countries for prioritising space programmes over the environmental quality of the planet they actually inhabit. However, in an ominous sign of the long-term instability that continues to mar the country he had renamed from Upper Volta, the 'African Che Guevara' was assassinated the following year, not long after his government had finished planting the 10 million trees he had envisaged there.[41] Only in 2005 was the baton picked up by two of Africa's most experienced and influential presidents, Nigeria's Olusegun Obasanjo, who advocated reforestation from coast to coast, and Senegal's Abdoulaye Wade, who gave Obasanjo's arboreal proposal a name.[42] Aware that this project would offer an unparalleled opportunity to connect eleven of its members both emotionally and physically, the African Union launched the Great Green Wall initiative two years later.

Almost immediately, the 'wall' encountered the same obstacle jeopardising its earlier counterparts in Algeria, the United States and China. What many rural Sahelian communities already knew, but the Great Green Wall's policymakers seemingly didn't, is that land degradation at the desert's edge is not owing to the desert itself; the latter only looks like it's advancing. Wade's rallying call that 'the desert is a spreading cancer' was a poor analogy for the Sahel's plight: although the imagery of sands marching menacingly towards green lands is visceral enough to capture people's attention, it is unhelpful in actually tackling desertification.[43] The truth is, trees are not well suited to being anti-desert barriers – not unless the soil and water supply are improved first, anyway.[44] Unfortunately, the initiative was forced to learn this lesson the hard way: northern Nigeria noticed that three-quarters of its 50 million new trees had died within just two months.[45] The original vision of connecting the Sahel with trees needed some significant tweaking, and quickly.

To succeed, local knowledge and expertise – which were largely or entirely overlooked in the first few years of the initiative – would have to constitute key building blocks. On top of the ecological challenges of growing trees on parched lands, it was implausible to expect time-poor communities to travel long distances to the remotest areas of the Sahel to tend to saplings, or to refrain from clearing trees whenever they needed fuel.[46] Cognisant of these issues, in place of a one-size-fits-all strategy of creating a coast-to-coast line of trees, the Great Green Wall has come to represent a necessarily varied patchwork of sustainable land and water management practices, in which communities are allowed to restore their lands in ways that best retain and respect their uniqueness, while serving their own fundamental needs.[47] It may be less obviously connective than its foreign-inspired roots, but in both its heterogeneity and its groundedness, the new Great Green Wall is more genuinely African.

Is it still a wall, though? Reflecting the symbolic power of lengthy, connective landmarks, which the Chinese tourist board knows only too well, the concept of a green wall is so irresistible that the name has remained in place even as the programme has evolved from a straightforward tree barrier into a far more comprehensive rural development initiative. In emphasising a collaborative approach to creating lush, productive landscapes that are distinct from one another and yet no less interconnected than the extensive belt of trees formerly envisioned, this strategy represents a reworking rather than a renunciation of the initial vision, a metaphorical 'wall' expressing shared resilience in the face of some of the world's most severe challenges. It doesn't matter that it won't look like a wall per se. As long as it is able to achieve the objectives of the original vision, devoid of 'holes' where land practices continue to exacerbate desertification, it will serve as a wall-in-kind, a quasi-natural entity representing a fitting boundary between cultivable lands and one of the planet's harshest natural landscapes. Ultimately, the benefits are all that really matter; the rest is just semantics.

The new and improved Great Green Wall exemplifies a growing determination among Sahelian communities to execute their own vision of a rejuvenated and interconnected landscape. In its updated format, the project tantalises in its sheer variety, bringing together an extraordinary array of re-greening ventures across the Sahel and increasingly beyond – for the Sahel's boundaries are ambiguous, and a more conceptual wall does not need to respect geography so stringently. Today, some of the countries involved in the initiative are not even close to what anybody would consider the Sahel, and instead act as partners in a more abstract sense, as in the case of Cape Verde, an island nation some 600 km (over 350 miles) off the African mainland, or Algeria, Tunisia, Libya and Egypt, all of which are to the north of the Sahara, not the south. Along with The Gambia, Ghana, Benin, Cameroon and Somalia (as well as Togo, indirectly),* almost the entirety of the northern half of Africa is now involved in the Great Green Wall, an expression of emotional and ideological affinity practically unrivalled in its scope. Certainly, the initiative hasn't forgotten its African architecture, and accordingly stresses that, regardless of whether a Great Green Wall programme was first envisaged by local famers or an international organisation, African ownership must be a prime precept. Considering that farmers haven't always been able to rely on foreign support and have instead been expected to adhere to foreign conceptions of a region so poorly known to most of the outside world, this principle is as pragmatic as it is ideological.

Each 'brick' of the wall has its own idiosyncrasies appropriate to local conditions and knowledge. In northern Burkina Faso, for example, a traditional agricultural technique for increasing soil fertility called *zaï* (elsewhere known as *tassa* or *towalen*) has been revived, enhanced and rolled out ever more widely since the early 1980s, thanks in no small part to the vision and energy of the late

---

* Since 2011, Togo has been involved in the Sahel and West Africa Program (SAWAP), which helps twelve African nations implement projects that explicitly support the Great Green Wall vision, but has not become a full partner of the initiative.

Burkinabé agronomist Yacouba Sawadogo. This method, which involves digging a grid of small pits into degraded soil and placing a couple of handfuls of organic matter at the bottom to attract termites, offers important advantages in regions where the soil is infertile, rainfall is sporadic and resources are scarce. In effect, the termites dig tunnels in the ground, allowing more rainwater to infiltrate into the soil, and process the organic matter so that nutrients in the soil become more readily available to any seeds carried into these moist pockets by wind, rain or hand.[48] With the endorsement of farmers cultivating significant yields of millet and sorghum – in turn prompting new tests on crops including maize, cotton, aubergines and watermelons – *zaï* has become an increasingly common feature of long stretches of the Great Green Wall's western segments, bringing new, verdant life to landscapes that only recently seemed beyond repair.

Other agricultural practices similarly manifest a frugal use of local materials, and despite their varied appearance, they regularly share the objective of capturing as much of nature's essential resources as possible. Often conjointly with *zaï*, and particularly where there is a slight gradient, many western Sahelian farmers choose to place rocks along natural contours to create long stony bunds or lines.[49] When the rain falls, these unbroken barriers slow surface run-off so that more water soaks into rather than rushing away from farmers' fields, while trapping valuable sediments and organic matter.[50] In Niger, another popular method on gentle slopes is to construct crescent-shaped bunds called *demi-lunes* ('half-moons'), capable of trapping surface run-off and organic matter immediately following rainfall.[51] In the clay plains of eastern Sudan, series of interconnected U-shaped earthen bunds, called *teras*, are traditionally used to prepare the soil for the cultivation of indigenous cereals such as sorghum (which can tolerate both drought and temporary waterlogging), while in Eritrea and Ethiopia, the historic practice of terrace farming is being expanded along these countries' slopes.[52]

In Chad, the prehistoric practice of forest gardening has been modernised as a regenerative agroforestry technique, whereby carefully selected trees are intentionally integrated with fruit and vegetable crops (for instance, aubergines, tomatoes and okra) to create a biodiverse and self-perpetuating ecosystem. With the significant support of the US non-profit organisation Trees for the Future, this strategy is, in effect, creating a green wall both edible and defensive, providing local communities with food and medicine, and providing protection for crops from wind and water erosion. Plus, instilling in local farmers the advantages of growing polycultures, underpinned by moister soil and a cooler microclimate, makes it increasingly likely that participating communities will enjoy greater food and income security as well.[53]

Most eye-catching of all are Senegal's *tolou keur*, community gardens comprising series of concentric circles. The outer halo of these gardens is made up of drought-resistant trees such as baobab and mahogany, protecting the smaller rings of food-producing (for instance, papaya and mango) and medicinal (such as aloe vera and sage) species within. Conceived by the Senegalese agricultural engineer Aly Ndiaye as a necessary response to declining imports, among other disruptions, during the COVID-19 pandemic, *tolou keur* are now regarded by local communities as essential laboratories for further experimentation as well as resource provision.[54] Along with curving permeable rock dams, round *bouli* ponds, strip-like rainwater trenches, V- and diamond-shaped micro-catchments, webs of live fences and field hedges, and parallel ploughed troughs, these human interventions are helping to shape distinctly geometric landscapes across the Sahel.

Thanks to the opportunities provided by the initiative for knowledge sharing across vast distances, various modern methods, including some conceived by people foreign to the Sahel, are also now being implemented throughout the region. Perhaps none is more noteworthy than farmer-managed natural regeneration (FMNR), which, having been successfully pioneered by the Australian agronomist, missionary and 'forest maker' Tony Rinaudo in Niger from the early 1980s, is today being replicated across larger and larger swathes of the western Sahel.[55] FMNR's rather inelegant name conceals its conceptual simplicity. Effectively, it requires farmers to take a relatively hands-off approach to their fields, intentionally not planting any trees and instead systematically pruning and nurturing those species that sprout spontaneously from the soil.[56] FMNR thus offers an ideal balance between the human and natural worlds, neither allowing the land to go fully wild, nor clearing or farming it to excess, all while enabling farmers to benefit from trees' ability to protect their crops from wind and wind-blown sand and to supply fodder for their livestock.

What this assortment of strategies brings to light is a growing appreciation that cutting-edge, high-tech but generally inflexible solutions requiring expensive maintenance are not the Sahel's reality.[57] Simple, low-cost improvements on locally familiar practices tend to be far more appropriate to the region's resource-scarce environment, with the very best options being those capable of addressing multiple needs simultaneously.[58] Methods that boost the soil's ability to retain water can allow farmers to grow larger and more varied species. Intercropping legumes with cereals not only augments the soil's available nitrogen (and, under certain conditions, phosphorus) but also provides multiple sources of income.[59] Fodder grasses such as *Brachiaria lata* are widely prioritised because they grow quickly and can be used as feed for animals or sold in markets as bales.[60] Tree species long overlooked by the authorities are now admired for their agroforestry advantages: for instance, the nitrogen-fixing apple-ring acacia *Faidherbia albida* unusually defoliates during the wet season, so it does not shade crops when they are best able to grow, and its fallen leaves help fertilise the soil.[61] Just as climate change, desertification and biodiversity loss operate as a vicious cycle trimming residents' incomes, land rehabilitation can increase groundwater recharge and support the development of farms and vegetable gardens capable of feeding stomachs a balanced and nutritious diet and wallets a steady supply of cash.[62]

If its reported statistics are to be believed, so far nowhere has been more adroit at rehabilitating its lands than Africa's second most populous country. At the beginning of the twentieth century, forests covered 40 per cent of Ethiopia (and 90 per cent of its highlands), but rampant deforestation, especially in the first half of the country's communist Derg dictatorship in the mid-1970s and early 1980s, left only 4 per cent of its lands forested by the turn of the millennium.[63] However, Ethiopia also has a long tradition of seeking symbiosis between the human, natural and spiritual worlds, represented best by its tens of thousands of church forests. Miniature Gardens of Eden, these ring-shaped woodlands, some of which are now safeguarded by stone walls, continue to be preserved as extensions of the sacred sites they surround, and in return offer congregants space for contemplation and protection from wind and heat.[64]

Considering Ethiopia's drastic transformation into a country many outsiders still associate with drought and famine, its church forests can appear like extraordinary remnants of a lush past, and yet the country's viridescent history may not be as lost as it once seemed. Aided, no doubt, by the country's recent issuance of landholding certificates to more than 360,000 households, public buy-in regarding its ongoing 'Green Legacy' initiative is quite astonishing.[65] By 2020, Ethiopia claimed to have restored 12 million hectares of degraded land (mostly outside the official Great Green Wall intervention zone), representing over two-thirds of the initiative's total to date, and putting the country 80 per cent of the way towards its own 2030 target.[66] Reflecting its inhabitants' deep-rooted reverence for trees, Ethiopia subsequently reported the planting of 25 billion seedlings between 2019 and 2022, and is currently striving to emulate this achievement by 2026.[67] Most remarkably of all, on 17 July 2023, millions of Ethiopians from all walks of life collaborated to plant more than 500 million trees, beating the country's previous record of 350 million, set four years earlier.[68] Though it's possible that the Ethiopian government has been a little cavalier with its data, when one considers that the country has simultaneously been facing insurgencies in three separate regions, any achievements even remotely close to these figures deserve to be congratulated rather than scorned.

Another major success story is a nation at the opposite end of the Great Green Wall, Senegal, where the planting of 12 million native trees in less than a decade seeks to protect crops from fierce gusts and citizens from economic insecurity.[69] Within its rejuvenated tree belt, perhaps no species better encapsulates this country's shrewdness than a thorny tree whose scientific name, *Senegalia senegal*, affirms its indigeneity and hence its natural survival advantages here. This acacia not only thrives, but also provides huge economic value: its gum arabic is used as an emulsifier in soft drinks and confectionery, as a base material in incense, as a thickener in shoe polish and as a binder for watercolour paints. The species is therefore almost as immune to deforestation as it is to drought, standing it among the likes of the African baobab *Adansonia digitata* (whose fruit contains an abundance of vitamin C and antioxidants), the desert date *Balanites aegyptiaca* (whose various parts are used for purposes as wide-ranging as food, medicine, board game pieces, jewellery and tattoo needles), and the so-called miracle tree *Moringa oleifera** (whose leaves are

---

* This species originated in Asia but has long been admired in Senegal for its resilience to harsh climate conditions and poor soil, a trait that has earned it the alternative nickname *nebeday* – a patois term meaning 'never die'. Another valuable tree in the western Sahel is *Vitellaria paradoxa* (whose shea butter is used in cosmetics, soap and cooking oil, among other things), while at the Great Green Wall's eastern end, two of the Magi's gifts to Jesus can be derived from species of *Boswellia* (frankincense) and *Commiphora* (myrrh).

often turned into powder and used as medicine or as an antioxidant-rich superfood) on a podium of regionally valuable trees.[70]

Senegal's relative success so far is buttressed by such sensitivity to local needs and preferences, as well as its eagerness to cultivate alliances with foreign partners, including, most peculiarly, the International Olympic Committee. Keen both to offset its substantial carbon footprint and to showcase Senegal's capital Dakar, which in 2026 will become the first African city to hold an Olympic sports event (the Summer Youth Olympics), the IOC is collaborating with the non-governmental organisation Tree Aid to grow an 'Olympic Forest' of nearly 600,000 native trees stretching from eastern Senegal to western Mali.[71] While this project has rightly drawn interest for its potential to benefit local communities economically and the wider world environmentally, its importance is also more symbolic. It shows how even though the original Great Green Wall plan needed to be replaced, there remains an enduring commitment to fashioning cross-border connections both palpable and ideological – a principle where the IOC has an obvious head start.

Having long been dismissed as the world's most hopeless region, here then is the Sahel emerging like a phoenix, resurrected from disaster through the foresight, knowhow and verve of African people. Whereas it has become commonplace for the rest of the world to view the Sahel through a lens both paternalistic and judgemental, stressing charity for its citizens while condemning its leaders (or otherwise ignoring the region entirely), the Great Green Wall in its modern iteration epitomises a growing recognition internationally that Sahelian people can take charge of their own affairs. The region's farmers are increasingly lauded for sharing their innovative cultivation methods with their counterparts throughout the region – people who, without European interference, might have been their fellow nationals. In so doing, they are, in effect, transcending the borders created by colonial powers in the past to carve up sub-Saharan Africa beyond recognition. In this sense, the Great Green Wall's significance is not merely environmental: it also represents a mutual, physical intermediary between people whose historic affiliations have been largely or entirely erased from the map. Meanwhile, by spreading the word about this 'world wonder' and pressuring their governments to back the initiative, citizens of countries beyond Africa are able to raise funds and build their own, ethically motivated kind of connection with the people of the Sahel, virtual in its manifestation and yet hardly less critical to the project's success. With a clear and shared vision for the region's future now in place, certain stretches of the Sahel are gradually being transformed into flourishing emerald landscapes, with a view to one day becoming interlocking stakes of this expanding palisade.

Across the globe, few initiatives can claim to be as inspirational as the Great Green Wall in attempting to unify broad and diverse societies with a shared, pressing purpose. In the region most severely affected by land degradation and desertification, where the soil is thin and rain is unreliable, here is a project whose mission is to grow not just plants, but the fertile ground on which they rely, too. In countries where extreme poverty is a fact of life for millions, this is an opportunity to create new economic opportunities, harnessing nature's resilience to glean value from its most essential resources. Terrorism and warfare, child marriage, forced marriage and female genital mutilation – these and other ills, disproportionately and deplorably common in the Sahel, may finally be eradicated by a project that seeks to empower the disenfranchised and create a better future for millions. Long marked by desperation and division, the Sahel may seem an unlikely place to find connection, but should the Great Green Wall mature and ripen as a fully integrated architecture of green and productive landscapes, as the uninterrupted green river from ocean to sea depicted on the initiative's official emblem, it will be living proof that humanity can work together to change itself and its environments for the better. More than just a symbol of African ingenuity, with the potential to alter how the continent is perceived internationally, and more than just a show of political will at the global scale, the Great Green Wall will be a proclamation to the rest of the world that the future never need be inevitable. If the world's most disadvantaged region can overcome its greatest threats, then surely anywhere can.

As testament to its international relevance, the Great Green Wall makes important contributions to a staggering fifteen of the United Nations' seventeen Sustainable Development Goals for a better world by 2030, all three of the conventions (on climate change, biological diversity and desertification) created by the renowned 1992 Earth Summit in Rio de Janeiro, and the United Nations Convention to Combat Desertification (UNCCD) and SDG 15's* shared objective of achieving Land Degradation Neutrality, again by 2030. It also acts as a logical flagship for the present UN Decade on Ecosystem Restoration and complements the African Forest Landscape Restoration Initiative's pledge to the Bonn Challenge, which seeks to restore 350 million hectares of degraded and deforested land by 2030. Already, the project has served as a green light for other endeavours whose mission is to advantage people and places with the biggest need. Impressed by the progress being made in parts of the Sahel, and conscious of the failures of its past decision to plant Aleppo pines (which demand lots of water, are susceptible to disease and have limited use for local communities) as the basis of its own green barrier, Algeria has opted to relaunch its afforestation project with far greater species diversity than before.[72] At the other end of the continent, plans to create a southern Great Green Wall are quickly taking shape, again with an emphasis on uncovering, rejuvenating and

---

* 'Protect, restore and promote sustainable use of terrestrial ecosystems, sustainably manage forests, combat desertification, and halt and reverse land degradation and halt biodiversity loss.'

sharing Indigenous farming techniques suited to drylands including the Kalahari and Namib deserts and South Africa's Karoo.[73] In stark contrast to the large-scale technical responses promoted half a century ago, which were far more successful in attracting publicity than in mitigating drought and famine, the methods typically preferred today tend to be couched in local knowledge and oriented to community priorities. This doesn't mean that they can't be shared internationally. It simply indicates that anti-desertification programmes are most effective when partners are modest enough to admit that no silver bullet exists, and instead work to find areas of common ground.

Indeed, when it comes to restoring degraded lands, collaboration among stakeholders, countries and projects is essential: one must not lose sight of the fact that improvements to water, soil and vegetation are mutually reinforcing, and act as the key to solving a broad range of economic, social and political issues. Accordingly, a multitude of initiatives in the Sahel have been directly harmonised with the Great Green Wall, from renewable energy to food security and from climate resilience to species conservation. For instance, the African Development Bank's ambition to create the world's largest solar zone in the Sahel, capable of supplying clean, renewable energy to a quarter of a billion people across the eleven original Great Green Wall countries by 2030, is explicitly tied to the initiative.[74] For one thing, by drastically reducing energy poverty throughout the Sahel, this Desert to Power Initiative ought to dissuade residents from cutting down the trees grown throughout the region for fuel. For another, through introducing solar-powered drip-irrigation systems, which enable farmers to grow more crops with less water and energy, it aims to increase agricultural productivity and food security – both key elements of the Great Green Wall's mission as well – while deterring people from practising less sustainable methods that threaten the wall's integrity. Rather than being a barricade against the desert, the Great Green Wall has become a bridge as much as a barrier, bringing together various people and places in the intricate and knotty fight against land degradation, ecosystem damage, climate change and extreme poverty and hunger. Backed by phenomenal political will on an international scale, the initiative's consequentiality, both real and symbolic, local and global, is hard to overstate.

Regrettably, though, bridges can be fragile things. Just as physical ones may be destroyed wilfully to sever essential connections, so too can more ideological bridges be impaired by all manner of causes, even to the extent that they are never completed. As a simultaneously tangible and conceptual entity, the Great Green Wall is doubly vulnerable, and although insufficient financing is an obvious obstacle to its eventual realisation, in many regions, political instability is more fundamental. Seeing a wave of coups d'état, armed conflicts (as well as the continuation of Senegal's persistent Casamance conflict) and ethnic violence erupt in almost every Sahelian nation in just the past decade, it is quite understandable that many potential international investors and donors have been reluctant to loosen their purse strings.[75] In a sick catch-22, without financial support, these countries are less and less able to develop the re-greening projects they need to coax despondent populations away from potential insurgencies and extremism. In the most acute cases – Djibo, for instance, a town in northern Burkina Faso that since February 2022 has been under blockade by armed rebels, forcing many of its inhabitants to survive on wild leaves – communities who want to transform their lands are unable even to access them.[76] Rather than becoming self-sufficient, over a tenth of this country's population has been internally displaced, and more than a fifth now depends on humanitarian aid, while the jihadist groups directing the siege of some four dozen locations level bridges and break pipes to slash local connections in the long run.[77]

At the same time, the fragility of the Sahel and Sahara's borders is enabling new kinds of international connections to develop, which are no less injurious in their impacts. Smuggling has become a particular problem in the Sahel, with Libya across the giant desert to the north representing an exceptionally abundant source of illegal weapons since the fall of its political strongman Muammar Gaddafi in 2011.[78] It is also more than a little curious that the United Arab Emirates imports significant amounts of gold from Libya as well as Togo, two African countries with very modest reserves, rather than Burkina Faso, whose bountiful mines have helped render it a key node on regional trafficking networks.[79] In this manner – and exposing how geographical connections can be co-opted and refashioned for alternative objectives – the trade routes that once brought prosperity to many in North and West Africa are now allowing wealth to evaporate from the area as fast as fallen rain on the Sahel's hard earth. Meanwhile, gold mines have become important focal points of instability in both Mali and Sudan, where Russia's Wagner Group mercenaries have siphoned money away from countries that desperately need it, to support Vladimir Putin's invasion of Ukraine instead.[80] In return, war has disrupted Russia and Ukraine's exports to the Sahel of wheat and maize, crops that are neither native to nor well adapted to growing in the region but that, along with rice, have become central to inhabitants' diets ever since famine in the 1960s left hundreds of thousands dependent on foreign food aid.[81]

In a geographical rejoinder to Newton's Third Law, what the unexpected relationship between the Sahel and Eastern Europe illustrates is that connection is never completely one-way – a fact we often notice only when the chain *breaks*. Starbucks, which purchases approximately 3 per cent of the world's entire coffee supply, is one US multinational keen to keep a close eye on the Sahel, principally Ethiopia. But of even greater intrigue are the potential complications facing soft drink consumers. Whereas Sudan rarely receives much attention on the news, as a major producer of gum arabic, its war has made at least two of the world's biggest corporations, Coca-Cola and PepsiCo, rather nervous about the supply of one of their essential ingredients.[82] Among those most affected by this specific logistical fissure are the Indigenous Tzotzil Maya people of San Juan Chamula in southern Mexico, who since the 1960s have used Coca-Cola in their ritual practices, believing that it has healing properties.[83] Recognising its unique popularity here, as well as the low cost of its labour force, Coca-Cola established a bottling plant nearby in 1994, which helps cater to this disproportionately large market while simultaneously causing major water shortages locally due to its significant water demands.[84] Bringing together a single product and a shared concern with drought (if not nutrition), the relationship between Sudan's 'gum belt' and the Chiapas Highlands is one of the world's most unusual and unexpected, and demonstrates distinctly how the Sahel's challenges can have significant ramifications far away.

Dire though this general situation of cross-border smuggling and conflict undoubtedly is, an even greater threat to the types of international connectivity sought by the Great Green Wall is the risk of the Sahel's bare earth becoming a ground zero for terrorism. Already there are signs of the Great Green Wall succumbing to the Sahel's political volatility. In 2023, the G5 Sahel security alliance collapsed due to severe instability in Mali, Burkina Faso and Niger, the three countries in the middle of this geographical chain; coups in these countries as well as Sudan have also led to all four countries being suspended from the African Union since mid-2021. To make matters worse, the remainder of the African Union now tends to view peace, security and governance as more vital concerns than the environment,[85] and somewhat surprisingly, omits the Great Green Wall from its fifteen 'Agenda 2063' flagship projects, 'key to accelerating Africa's economic growth and development as well as promoting our common identity'.[86] Declining accountability tends to follow a loss of commitment, and with a 'quantum leap' required to raise far more money than the approximately US $1.8 billion allocated to the project by its halfway stage (for reference, in 2020 the UNCCD estimated that $3.6–4.3 billion annually would be needed over the next decade for land rehabilitation measures alone),[87] it is understandable that significant doubt has set in that the wall will ever be completed.[88] Building a continuous chain of restored land across the Sahel had always seemed a bold ambition, but few stakeholders seemed to predict quite how challenging this task would prove to be.

A rare but necessary example of the contemporary Sahelian nations working together, the Great Green Wall uses the power of connection to mitigate and reverse this vast belt's most formidable perils. The problem is that the same qualities that provide cause for hope and which might hold the endeavour (and by extension, the region) together also risk tearing it apart. The initiative, in short, is characterised by paradoxes that threaten its eventual realisation.

First is the fundamental but thorny matter of whether the Great Green Wall is a top-down initiative overseen by a central body (in line with the original tree-barrier vision), or a genuinely bottom-up enterprise led by different local communities (as implied by the new concept). In reality, the project now resembles something of a mishmash, being nominally coordinated by the Pan-African Agency of the Great Green Wall (PAAGGW), a Mauritanian-based regional authority that finds itself regularly bypassed by funders keen to control how their contributions are spent.[89] Before one considers the qualms a smaller organisation may have in contributing funds to countries still widely stereotyped as corrupt, even the European Union and World Bank have tended to provide their money directly via individual governments rather than through the PAAGGW.[90] Unsurprisingly, this agency and the African Union have therefore struggled to determine how much money is available, who is funding what, and whether the finances represent loans, grants or otherwise, the conditions for which vary greatly. To complicate matters further, the Sahelian countries are often now hesitant to accept money in the form of loans, particularly for environmental projects, fearing that these will only compound their ceaseless and traumatic debt crises. Together with a relative lack of independent research tracking how much funding each country has mobilised and the progress it has made towards its targets (how many planted trees have actually survived, for instance), the question of who does or should lead the initiative risks frustrating its future.[91]

An associated paradox pertains to donors. While the initiative is directly relevant to a wide range of environmental and development objectives, due to its sheer scale and scope – not to mention a paucity of centralised information about specific Great Green Wall programmes – prospective donors often struggle to determine where and how they should help.[92] Relatedly, even though the Great Green Wall's emphasis on community participation ought to encourage local populations to contribute, because it comprises numerous smaller projects necessarily distinctive to local contexts, there is a real risk that many will be treated as discrete silos or will remain out of the limelight necessary for attracting wider investment. So far, Ethiopia and Nigeria have been among the most proactive partner countries in obtaining funds and implementing new projects,[93] but owing to the considerable strain applied by their rapidly growing populations across significant portions of territory, they are also under the most pressure to continue attracting financial support, certainly compared to far smaller nations such as Djibouti. Size commands the attention of funders, but it demands that they spend more as well.

A further challenge with managing such a colossal project simultaneously from the grassroots and from above is ensuring close alignment between parties. The problem is that despite the theoretical flexibility of the Great Green Wall as an umbrella initiative, national land-use laws, particularly with regard to tree felling, still tend to demand burdensome and slow bureaucratic procedures, which can sap enthusiasm for the initiative. In much of the western Sahel, the mixed legacy of customary law and colonialism means that agricultural and pastoral lands continue to be owned by the state.[94] Additionally, rural residents here can find it far harder to obtain legal land title to lands their families have managed for generations than do foreign investors and loggers, whose interests tend to be narrower and whose actions are more likely to be exploitative.[95] As a consequence, many communities are understandably reluctant to engage in sustainable practices such as FMNR or even to grow trees in the first place: not only do they see little incentive to manage trees, but doing so also risks punishment from powerful figures already liable to treat them with suspicion.[96] It doesn't matter that globally, Indigenous people are incomparable stewards of the environment, protecting as much as 85 per cent of the planet's biodiversity despite representing less than 5 per cent of its population.[97] In the Sahel, the established logic is that the authorities know best. Yet without local knowledge, engagement and collaboration from West to East Africa, there will never be a Great Green Wall. The question of whether the Great Green Wall is a truly unified initiative or a miscellany of small-scale schemes looms large again: oversight from above is crucial to integrating diverse projects as a continuous strip of re-greened land, but if local residents are unable to take ownership of their own needs and undertake practices that make sense to them, there will be little to connect.

This raises a related puzzle: although piecemeal projects are easier to implement, only a fully integrated programme can overcome the Sahel's combined environmental, economic, social and political objectives. Cross-border collaboration is essential to the Sahel just as contiguity is critical to the Great Green Wall's being, for the region and, by extension, the wall is only as strong as its weakest link. Land degradation in one place immediately puts strain on other sections; too much and the risk of instability escalates and spreads. Much the same can be said of extremism, which additionally deters investors from supporting those places in the direst need.

On the flipside, a solution in one location can also have

unanticipated knock-on effects elsewhere, bolstering the need to consider how seemingly distinct places interact and are connected as a single, consolidated, cross-border system. Every question, every decision, needs to be considered in relation to everything else. What value does a foreign-funded project have if it is only short term and fails to inspire the locals who will be handling it for years to come? What point is there in water harvesting if grazing controls aren't implemented, or in producing crops if no viable market exists? And how will radicalism ever be tackled effectively as long as civilians are treated as potential opponents of the state and its interests? As long as the wall has cracks, it's effectively just a loosely affiliated collection of scattered gardens, beneficial to local residents in very select areas but unlikely to bring lasting peace further afield. And at the time of writing, the existence of mere cracks would be regarded as a triumph. Africa as a whole saw a far larger net loss of forest area than any other part of the world between 2010 and 2020; and as of 2023, the Great Green Wall remained just one-fifth of the way towards its land restoration target.[98] (Incidentally, an associated dilemma is that the more the wall grows, the more vulnerable long sections of it become to being devoured by desert locusts unable to believe their luck at the development of the world's largest buffet table.)

Still, the most important question is whether this new Great Green Wall is really a wall at all. It certainly offers a lot more advantages than the original tree-barrier vision. Given that each community can now define the Great Green Wall as it sees fit, the project has become more inclusive of local preferences and contextual needs, and hence more likely to be realised. But to a potential investor, the new vision lacks the very thing the original offered in abundance: an immediately comprehensible image of connectivity. A long, tangible tree wall is an exciting prospect, a physical monument of transcontinental connection. While it's true that a mosaic of restored drylands is infinitely more practical and is capable of integrating a far broader area of sub-Saharan Africa than a single strip of re-greened land, because the connection it offers is so much subtler, it is less able to inspire international elites and donors, who tend to prefer more straightforward projects.[99] As the case of so many other walls throughout history shows, people tend to favour the simplest, seemingly most clear-cut options rather than complex yet effective solutions.

Recognising some of these impediments, in January 2021 the fourth edition of the One Planet Summit (which brought together government officials, entrepreneurs, international organisation heads and other leaders concerned with protecting the planet's biodiversity)* saw the pledging of billions of dollars of new financing and the launch of an 'Accelerator' to clarify types of funding commitment and better coordinate, monitor and track the Great Green Wall's implementation. Considering the obvious international scope of the project and the associated tempta-tion to put more power in authorities' hands, the Accelerator's focus is reassuringly and necessarily oriented to the grassroots, stressing the importance of increasing agricultural communities' resilience to environmental challenges, improving food security and job opportunities, and supporting participating countries in monitoring their specific projects and directing funds towards the biggest needs.[100] In this sense, the Accelerator seeks to achieve the optimal balance between enabling local communities to address those issues that afflict them most directly and ensuring that enough oversight exists to ensure that each country is kept on the same page.

Such principles – adherence to which has since been supported by the creation of a digital 'Observatory' platform (in 2024) to monitor progress – are essential for building accountability and bringing this transnational initiative to fruition.[101] Yet in spite of some important successes, the Great Green Wall continues to be

---

* The US, Russian, Indian and Brazilian head honchos were conspicuous by their absence.

given startlingly little attention in most Sahelian countries' environmental programmes and budgets.[102] Instead of being regarded as an organic yet holistic environmental-social-economic-political endeavour, the project is often still treated as a mere reforestation programme, generating discord where one groups feels that its own needs are being overlooked.[103] Why transhumant pastoralists are suddenly required to respect new fences protecting the Great Green Wall's saplings when herding their cattle along their traditional livestock corridors, or why nurseries often seem to be supplied with water faster than villagers, for example, can go unexplained and fuel new conflicts among people whose principal concern – creating a livelihood from little – is the same.[104] Furthermore, many of the most vulnerable members of society complain that the initiative ignores or excludes them: women with absent husbands, for example, who can find that cash-to-work programmes are gatekept by male community leaders and thereby entrench their social isolation, while their peers are able to collaborate and make an income.[105] Disappointingly, too, the initiative's international partners have often chosen to prioritise projects in more advantaged locations where their objectives can be realised most easily (for instance, by virtue of being more humid, or better connected to existing roads and markets), as opposed to focusing attention on the places that need help most.[106] Rather than bringing the region closer together, the project, in many ways, has magnified the Sahel's social and geographical inequities.

This is not to understate the Great Green Wall's laudable achievements. Over time, the landscape of much of the Sahel is becoming increasingly heterogeneous and complex, its fields boasting a growing diversity of flora while its pastures accommodate larger herds of cattle, goats and sheep.[107] Even though plenty of young adults still believe, not without valid reason, that their job prospects are superior in sub-Saharan Africa's swelling cities, a good many others are increasingly willing to stay and work together to bring life to injured plots.[108] A new 'ecopreneur' class has emerged among women, who, instead of spending significant portions of the day collecting wood for fuel, can tend to community gardens and manage shops.[109] More and more children enjoy diverse, nutritious diets, setting them up for success at school.[110] These accomplishments all provide real hope that the world's most disadvantaged region is sprouting a better future for its inhabitants. The challenge is to close the wall's many remaining gaps. New support from the Accelerator may have helped the initiative pick up the pace, but billions more dollars are still needed to ensure it doesn't remain a mirage at the desert's edge.

Beyond its potential to improve the lives of millions of people across the Sahel, and beyond its showcasing of sensible land-use strategies that organisations can disseminate elsewhere, there are important reasons why the rest of the world should care about the Great Green Wall. At the same time as boats of despondent African migrants make the perilous journey across the Mediterranean, this project potentially provides a convincing incentive to stay. It is no accident that the European Union, keeping at least half an eye on the continent's future stability, has invested millions of dollars into the initiative, while France, home to easily the largest population of African descent in Europe and typically interventionist in its counter-terrorist military operations in the Sahel, is one of its biggest champions.[111] As many Europeans and even North Americans periodically notice when the sky turns an eerie orange, sand particles accumulate on vehicles and respiratory problems (among other health concerns) spike, the Sahara is easily the world's biggest source of wind-blown dust, so any initiative to tackle land degradation in Africa will ameliorate life on other continents as well.

Walls are often viewed as forbidding entities conveying separation and a lack of welcome. Those on one side may particularly wish to destroy or overcome such a structure, in bitter disagreement with their opponents on the other. This project has no such patina. Indeed, it proves that such entities can provide cause to stick around and work together to improve the future for all. It proves that walls don't need to be built out of paranoia, and they don't need to isolate deprecated communities. Instead, they can provide insulation against real issues, both human and natural, which in one way or another affect us all (and for which, through our consumption habits and energy use, we are in turn at least partly responsible). The geographical connections we shape are generally tangible, but in certain circumstances, they can exist more conceptually. If completed as planned, the Great Green Wall will, in essence, be both: a symbol of hope as well as a vehicle of peace and prosperity, protective and productive all at once.

7

Co-option: Chicago's ridges and waterways

What right had these people to our village, and our fields which the Great Spirit had given us to live upon? My reason teaches me that land cannot be sold … Nothing can be sold but such things as can be carried away.

Black Hawk[1]

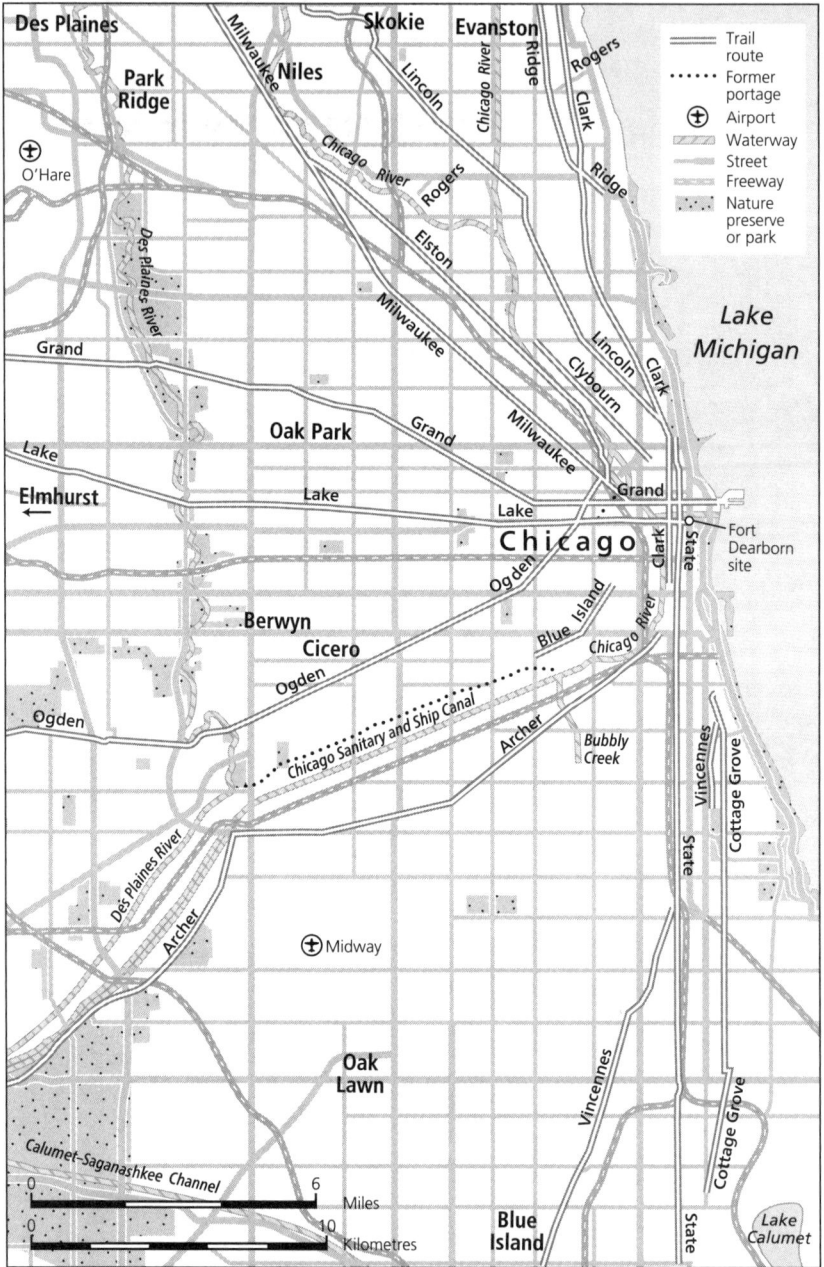

Des Plaines

Park
Ridge

Skokie

Niles

Evanston

Milwaukee

Lincoln

Chicago River

Ridge

Rogers

Clark

⊕
O'Hare

Chicago River

Rogers

Des Plaines River

Elston

Milwaukee

Grand

Oak Park

Grand

Milwaukee

Lake
Michigan

Lincoln

Clybourn

Clark

Lake

Elmhurst ←

Lake

Lake

Grand

Ogden

Chicago

Clark

O State

Fort
Dearborn
site

Berwyn

Cicero

Ogden

Ogden

Ogden

Chicago Sanitary and Ship Canal

Blue Island

Archer

Chicago River

Bubbly
Creek

State

Vincennes

Cottage Grove

Des Plaines River

Archer

⊕ Midway

Oak
Lawn

Vincennes

Cottage Grove

Calumet-Saganashkee Channel

Archer

0   6
Miles

0   10
Kilometres

Blue
Island

State

Lake
Calumet

Trail
route

Former
portage

⊕ Airport

Waterway

Street

Freeway

Nature
preserve
or park

Humour me, if you will, with a quick trivia question. What do the following all have in common?

- Chocolate brownies
- Mobile phones
- Softball
- House music
- TV remote controls
- Skyscrapers
- Spray paint
- Playboy Bunnies
- Car radios
- Firemen's poles
- Soap operas
- Ferris wheels

Congratulations if you knew – or at least suspected from the chapter title – that each originated in Chicago! Chicagoans have long loved to boast about their city's miscellaneous contributions to the world, so much so that its most popular nickname, 'the Windy City', refers not to its lakeshore gusts but to the widely observed penchant for bragging among its brash, 'windbag' politicians and citizens.[2] In fact, the list of Chicagoland* inventions is arguably even longer and more varied, with some contending,

---

*Chicagoland (that is, the metropolitan area) comprises Chicago, as well as its suburbs and satellite cities.

not without good evidence, that the first zip, hospital blood bank, vacuum cleaner, ice-cream sundae and atomic bomb were at least partly developed there.

Chicago was built on steel, rail and meatpacking, and even as its economic base has gradually diversified to include flourishing leisure and entertainment sectors (among others), it remains deeply proud of its industrial history and its citizens' incessant desire to innovate and inspire. It is a city of 'big shoulders'[3] and bigger ambitions, where 'dragged through the garden' hot dogs and deep-dish pizza were designed to sustain hungry workers, and where protests such as the Haymarket Affair (1886) and the Pullman Strike (1894) ultimately helped the country's trade unions achieve both a 'nine to five' workday and a national holiday.[4] In the twentieth century, Chicagoans ranging from Ernest Hemingway to Walt Disney entered the pantheon of modern cultural icons, while the Michael Jordan-era Bulls generated sensational interest in the National Basketball Association far beyond the United States' borders. With Chicago's improv comedians and blues musicians, gourmet chefs and world-class architects influencing millions across the planet to the present day, it is surely not controversial to state that modern life would look very different without Chicagoans' motley endeavours and inventions over the generations. But lost among impressive lists of what Chicago has given society is what Chicago has taken from others. For no contemporary city was founded on a blank slate – and Chicago, which was incorporated as a city as late as 1837, is younger than most. Whose voices have been hushed, whose expertise appropriated, whose world upturned, all in the name of modernisation? How has Chicago and the land on which it stands so proudly been shaped and reshaped to suit differing and shifting perspectives, needs and objectives?

Consulting a map of Chicago for the first time, one would be hard pressed not to notice the city's precise grid system: almost all its streets run either north to south or west to east, with eight blocks consistently equalling one mile.* Viewed from the top of the Willis Tower or 875 North Michigan Avenue (which staunch Chicagoans still refer to by their former names, the Sears Tower and the John Hancock, respectively), the grid becomes even more obvious, with wide boulevards and avenues stretching without a single deviation across the plains into the horizon as far as the eye can see. Arising in the morning, a visitor at ground level may find themselves blinded by the sunrise as they look east towards Lake Michigan, such is the city's near-flawless alignment with the cardinal directions of a compass. And should they be in the downtown 'Loop' on either equinox, they may be fortunate enough to witness the magnificent 'Chicagohenge', when the rising and setting sun beams brilliantly between the skyscrapers lining the city's east–west streets, turning the sky a fiery orange.

However, upon closer scrutiny, Chicago's map presents a few oddities. The freeways, although somewhat loyal to Chicago's straight-line dogmatism (even if Interstates 90 and 55 run diagonally rather than north–south or east–west), deviate on occasion, perhaps most eye-catchingly 2½ km (roughly 1½ miles) north-west of the Loop, where the Kennedy Expressway bends around the grandiose St John Cantius Catholic Church. The twisting North Branch of the Chicago River forces a small number of streets to diverge from the grid, even if the main thoroughfares tend to continue running true and roughshod across the landscape. And approximately following the configuration of the shoreline and its band of connected parks, Lake Shore Drive snakes from north to south with an eccentricity befitting its abbreviation, 'LSD'. Still, most interesting of all are the dozen or so roads that appear entirely disinterested in adhering to the lattice, instead spreading

---

*Furthermore, because address numbers increase by 100 moving away from the centre (specifically the intersection of Madison Street and State Street), calculating distances here is relatively straightforward. For instance, a 4000 North address is forty blocks north of Madison Street, or five miles.

outwards from the Loop as winding diagonals. While their names honouring Midwestern settlements (Milwaukee, Vincennes, Green Bay) and luminaries (Lincoln, Clark, Ogden) may suggest they are little different from other Chicago streets, their unconventional configuration connotes a more distinctive history, rooted in a physical geography refashioned beyond recognition. For, having been nurtured by Indigenous groups for millennia, the region's natural landscapes were seized and their innate advantages co-opted according to an alternative, interventionist precept of how people might engage with the planet. Chicago's few sinuous streets offer a rare glimpse into a past otherwise concealed beneath the city's systematic and polished veneer. They, along with the city's atypical waterway to the south-west, entreat us to dig more deeply, so that the extraordinary stories associated with its Indigenous past can rise to the surface.

With a population of 9.5 million spread across nearly 30,000 km² (over 10,000 square miles), Chicagoland today would be completely alien to a person living in or travelling through this marshy area as recently as 200 years ago. For thousands of years, Indigenous groups including but certainly not limited to the Potawatomi, Ojibwe, Odawa, Illinois, Miami, Sauk and Meskwaki* travelled along narrow ridges of unconsolidated debris which had been deposited by retreating glaciers to converge at this point.[5] Despite their unassuming nature, these ridges, which generally measured just 1.5 metres (5 feet) wide and rarely rose more than 3–4 metres (12 feet) above the general terrain, allowed these groups to cut across the poorly drained wetlands and muddy prairies at the south-west corner of Lake Michigan.[6]

Until the Potawatomi founded what is believed to have been the

---

* Alternative names for these groups are, respectively, Bodéwadmik, Chippewa, Ottawa, Illiniwek, Myaamiaki, Asâkîwaki or Sac, and Meshkwahkîha or Fox.

first permanent village on the Chicago River in the mid-eighteenth century – which was quickly joined by Odawa and Ojibwe communities as well – settlement tended to be temporary.[7] Instead of attempting to bend the natural environment to human needs by establishing fixed farms and hunting bases, Indigenous communities saw that humans and other living beings all occupy the same, unified system.[8] Guided by this spiritual understanding of people's place in the world, while recognising that the region's baking hot summers and intensely cold winters obligated more than a small degree of adaptation on humans' part, the convention was to rotate among various favoured locations, hunting, fishing and growing wheat and corn wherever and whenever the conditions were right.[9] Thus human presence on the land was necessarily sustainable; conscious stewardship of the environment allowed communities to maximise their yields without exploiting nature's gifts to excess.[10] Although occasional warfare in the eighteenth century, often involving British or French troops, was another factor that had dissuaded long-term habitation, the sinewy ridges over which Chicago now sits ensured that the site continued to see fairly consistent foot traffic.[11]

What's more, the area was not only advantageous because of its ridges. Thanks to the configuration of Lake Michigan and the spidery rivers to its west, much of the general region is well suited to portaging, whereby a person only needs to carry or haul their canoe a fairly short distance across marshy ground to reach a new body of water.[12] By effectively attaching otherwise disconnected rivers and lakes, portages represented key links in Indigenous groups' transport networks – and none was more helpful than the Chicago Portage between the Des Plaines River to the west and the Chicago River to the east. Because this location stands on the St Lawrence River Divide, separating those rivers that flow into the Great Lakes and the St Lawrence River from those such as the Mississippi that run towards the Gulf of Mexico, a person with a paddle could travel as far as

Canada's Atlantic coastline, the warm southern states or the Rocky Mountains, with no or minimal walking required.[13] As a consequence, generations of Indigenous communities from across much of the continent were able to converge at this recognised node, and trade an unparalleled assortment of necessities and treasures.[14] There was simply no impetus to control or reshape the area's physical geography; adaptation was the responsibility of humans.

For a time, newcomers agreed with this worldview, or at least acquiesced: the earliest European visitors quickly learned from the region's Indigenous communities how best to navigate its waterways, not to mention survive in a place where winter temperatures frequently fall well below freezing. The first documented Europeans to reach Chicago were a French expedition in 1673 headed by a Jesuit missionary named Father Jacques Marquette and a fur trader called Louis Jolliet, part of whose assignment was to confirm rumours of the great Mississippi River and from there to explore whether it led, as they mistakenly believed, all the way to the Pacific Ocean.[15] Having been shown a portage in Wisconsin (near the contemporary city of Portage) by members of the Miami community, Marquette and Jolliet were able to travel not west but south along the hallowed river as far as Arkansas. Only here did they turn around, on the advice of a group of Michigamea that Spanish colonists had recently been seen in the area and would likely attack a European rival. It was on this return journey that the French team encountered the 11-km (7-mile) Chicago Portage, thanks to a Kaskaskia tip that even despite its seasonally fluctuating water levels, this route offered quicker and easier access back to Lake Michigan.[16]

With 6,000 km (3,750 miles) under their belt and new alliances in hand, Marquette and Jolliet's odyssey proved to be a resounding success for France. Their notes and maps gave their country profound knowledge of the continental interior and offered insights into how it might – and indeed would – soon establish colonies and settlements along the Mississippi, connecting New Orleans to the Great Lakes and beyond.[17] Jolliet even posited that a canal could be built at the Chicago Portage to ease travel for their compatriots in future, pre-empting the eventual construction of the Illinois and Michigan Canal through this area in 1848.[18] Meanwhile, the region's Indigenous communities had been given little reason not to trust Europeans thus far. In particular, Marquette's proficiency in several Indigenous languages and dialects (most notably Miami-Illinois), as well as his willingness to acquaint himself with and participate in local customs such as smoking sacred calumet pipes, suggested that other visitors would be similarly genial and appreciative.[19]

Such qualities were certainly true of the Chicago area's first permanent, non-Indigenous settler and therefore the city's unofficial founder. Jean-Baptiste Point DuSable, a man whose early life remains mysterious beyond the fact he spoke French and was of African descent (possibly he was born in Sainte-Domingue, now Haiti), married a Potawatomi woman named Kitihawa and built a homestead and trading post near the mouth of the Chicago River in the late 1770s. Not merely successful farmers and entrepreneurs but diplomats too, the multilingual DuSables were fundamental to maintaining the peace whenever a dispute arose between the various people and groups passing through the area.[20] Through their relationship and their actions, they were a living embodiment of the idea that different blood need not preclude reciprocity – that in fact, diversity could enable collaboration and mutual understanding.

However, Indigenous networks soon started to come under threat. In 1785, the nascent United States brought an entirely new way of understanding and engaging with hitherto 'undeveloped' western lands, whereby they would be subdivided into regular square plots each measuring one square mile (640 acres), ready to be sold to private citizens. Two years later, the lands bounded by the Great Lakes, the state of Pennsylvania and the Mississippi and

Ohio rivers were incorporated as the country's first new territory, the unimaginatively named Northwest Territory. Although from a conceptual point of view this region's boundaries corresponded well with what many today regard as the Midwest, incorporation would prove to have far broader and tangible impacts on the country's evolution. First, procedures now existed for settling and organising Indigenous lands to the west, with the eventual goal of admitting new states, where slavery would be prohibited, to the union.[21] And second, Indigenous communities' cooperative, reciprocal relationship with the natural environment started to be supplanted by an unfamiliarly human-centric approach of uniformly carving up the land for the purposes of permanent settlement and profit-making.[22]

Even so, Chicago's emergence as a meticulously planned grid city was far from immediate; only in 1830 was the first plat (map) of proposed lots created by a quadrilateral-devoted surveyor named James Thompson for what is now a small portion of downtown Chicago.[23] In the first instance, the communities obstructing a particular kind of American dream needed to be expelled. Still short of money after the Revolutionary War (1775–83), the young US government hoped to raise revenue by selling plots of the Northwest Territory, a task complicated not only by the continued presence of Indigenous groups, but that of British forces, too, who remained determined to create an Indigenous buffer state capable of blocking the new country's westward expansion, while also protecting British fur trade interests.[24] Following a ten-year war between the United States and a confederation of Indigenous groups backed militarily and financially by Britain, a treaty was signed at Greenville, Ohio in 1795 to redefine the boundary between Indigenous and white settler lands.[25] Although the treaty mostly pertained to Ohio, it crucially handed a small portion of land at the mouth of the Chicago River to the United States, and permitted the latter to establish an army post at the site. Within the decade, the United

States had planted its westernmost garrison here, Fort Dearborn, directly opposite the DuSables' former homestead.

Following further territorial disputes between various Indigenous communities on the one hand and American settlers, hunters and surveyors on the other, a series of treaties was subsequently signed near St Louis between 1804 and 1825 to compensate the former for a strip of land between Lake Michigan and the Mississippi River, thereby allowing the United States to spread further westwards, including along the Chicago Portage. Whether the Indigenous representatives fully understood the terms of the treaties is very questionable. In addition to the fact that many signed them with an 'X', the settler concept of private property ownership (not to mention different types of property rights) was incongruous with most groups' tradition of sharing the land among the community.[26] At least as confusing was the fact that the 1816 treaty permitted Potawatomi, Odawa and Ojibwe people to 'hunt and fish within the limits of the land hereby relinquished and ceded, so long as it may continue to be the property of the United States', thus authorising Indigenous presence as long as they officially handed over the same territory.[27] Now the new land had been acquired, the planning of a brand-new canal between Lake Michigan and the Illinois River commenced, consistent with Jolliet's earlier recommendation.

Rightly fearing their ways of life were being threatened, many Indigenous communities continued to fight back, still often in collaboration with lingering British troops who harboured hopes of reclaiming the American interior. Tensions came to a head in 1812, with a Potawatomi assault on Fort Dearborn. To many Americans, this event constituted a massacre and an atrocity, as scores of troops and their families were rounded up and killed by Indigenous warriors, while others were ransomed; to others, however, including many Indigenous people, the events at Fort Dearborn constituted a Hail Mary, a desperate act of self-defence against a powerful adversary already determined to squeeze them out.[28] Although the establishment of Fort Dearborn is today represented as one of four red stars on Chicago's flag* and its approximate location is marked by a plaque and bronze inserts in the pavement, both emblematic of the settler perspective – Chicago's creation story – the city's collective memory became somewhat nuanced in 2009 when a small park was given the more neutral name 'Battle of Fort Dearborn Park'.

Nevertheless, in the short term the attack on Fort Dearborn only invigorated a far more partisan response. Having finally dislodged the British via an 1815 peace treaty (as well as rebuilding the fort in 1816),† the US government committed itself to an increasingly active policy of dispossession, especially in areas already inhabited by white settlers.[29] Greedily eyeing the 2 million km² (over 828,000 square miles) of territory purchased from cash-strapped France in 1803, stretching from southern Alberta in what is now Canada to the Bayou region on Louisiana's Gulf Coast, a rapidly growing body of public and political figures peddled expansion as a moral imperative, a God-given mission by which the US would control and profit from western lands.[30] In the minds of many speculators and entrepreneurs, already excited by the outcomes of Lewis and Clark's reconnaissance expedition to the Pacific Northwest of 1804–6, Indigenous people were wasting, not maximising, the sheer potential of the vast, mineral-rich West. Particularly perplexing to many white politicians and settlers were Indigenous groups that opted to live nomadic or semi-nomadic lives, rather than establishing permanent homes

---

* The other three stars also reference important events in Chicago's history: the Great Chicago Fire of 1871, and the city's hosting of two world's fairs in 1893 and 1933–4.

† Interestingly, the three-year war between the United States, the United Kingdom and their respective Indigenous allies saw the first and only foreign occupation of the US capital since the Revolutionary War, when British troops set fire to Washington DC, including the US Capitol and the White House, in August 1814.

and fences to claim and manipulate specific portions according to their individual needs.[31]

As scepticism grew that Indigenous people could ever be effectively assimilated into white America and new precious metal reserves began to be discovered even in 'old' lands, 'removal' became the watchword.[32] Dissatisfied by his predecessors' comparative hesitancy, on ascending to the presidency Andrew Jackson quickly ramped up the pressure on Congress, signing into law the 1830 Indian Removal Act to forcibly relocate all Indigenous communities living in a US state or territory west of the Mississippi River.[33] By this time, white paranoia about Indigenous people had become so intense that when the Sauk leader Black Hawk and his followers returned to Illinois in 1832, hoping to plant crops as they had been doing for decades, 7,000 US soldiers were deployed to hunt down Indigenous people until they could feel satisfied that white settlements would face no further resistance.[34] In 1833, a month after Chicago was officially incorporated as a town of 300 people, a final treaty was signed obligating Chicago's Potawatomi,* Odawa and Ojibwe peoples to cede their remaining 5 million acres (more than 20,000 km²/nearly 8,000 square miles) in the Great Lakes region to the government, and resettle 800 km (500 miles) south-west in the unincorporated territory of Kansas within three years. As they departed in 1835, several hundred warriors performed their last dance on Chicago soil, an act of defiance against a country and a settlement that had betrayed them.[35]

From this point onwards, the remote, swampy military and farming outpost morphed into a modern metropolis practically

---

* The only exception was four Pokagon Potawatomi village communities, whose chief Leopold Pokagon managed to negotiate an amendment to allow them to stay in Michigan on account of the fact he had previously converted to Catholicism to prove that his people were sufficiently 'civilised' to live among white Americans.

overnight. Most discernibly, Chicago's population exploded. Whereas just 4,000 people lived in Chicago when it graduated from 'town' to 'city' in 1837, by 1890 more than a million people called it home. Within the next twenty years, this figure doubled to 2 million.[36] No other city in the world grew as quickly in the same time span. Alongside this expansion, the settler mentality that drove it matured from simply divvying up the land presented by nature, to manipulating nature so that land and water would better serve humans' evolving demands.

Rapidly, the city set about thoroughly transforming the surroundings Indigenous communities had respected and cherished since time immemorial. The Illinois and Michigan Canal, which provided a direct connection between Lake Michigan and the Illinois River via the Chicago River, was the first of several major projects. Dug entirely by hand, mostly by desperate, destitute Irish immigrants, the canal opened in 1848. This sounded the death knell of the Chicago Portage, replacing a natural but marshy and hence suboptimal transport link with a modern waterway capable of fostering large-scale trade between the East Coast and the Gulf of Mexico.[37] Whereas previously boats struggled to contend with the seasonally variable terrain of this 'Mud Lake' – if they could even enter it from Lake Michigan, where a wide, protruding and seemingly indestructible sandbar at the river mouth had long prevented access by more sizeable vessels – now ships could carry bulky loads from waterway to waterway without needing to stop.[38] Not merely relevant to the increasingly populous Chicago area, the canal made it possible to voyage between the country's biggest and third biggest cities, New York and New Orleans, supporting the growth of all three as well as numerous new settlements on the main route through the continental interior.*

Meanwhile, human hubris soon led to new problems that

---

* Incidentally, had it not been for the canal, Chicagoans would most likely have a Wisconsin rather than an Illinois address: on becoming a state in 1818,

required additional technological corrections to Chicago's natural environment. Cholera had afflicted Chicago as early as the 1830s, having been introduced by the soldiers brought in to fight Black Hawk, but its relationship to contaminated water wasn't understood until the mid-1850s. By this time, the metropolis on the marsh was suffering from frequent and severe epidemics of both cholera and dysentery, and simple sewers would not suffice, for the high water table rendered digging futile even after the surface water had been drained to permit new developments. The solution was astonishing: the city would be physically raised as much as 4 metres (14 feet) on jacks to accommodate sewers underneath; buildings that could not be lifted were placed on rollers and moved elsewhere, often while workers proceeded with their daily tasks inside.[39] This legacy can still be seen in some older neighbourhoods, where the entrances to many houses are on the floor above ground level, often via a short flight of stairs or a drawbridge, because the original owner was unable to afford to raise the building to the new street level. Passing by construction sites, perceptive pedestrians might also notice an empty space beneath the current and former levels of the street and pavement.

Unfortunately, water-borne disease outbreaks continued regardless. A major waste management faux pas had been made: the sewage system discharged all kinds of waste straight into the sluggish Chicago River, which flowed naturally into Lake Michigan. No prizes for guessing the source of Chicago's drinking water. Particularly nauseating was the dreck from the Union Stock Yards, from 1865 the world's biggest meatpacking district and for a time a perhaps bizarrely popular tourist attraction.[40] Here, so many animal carcasses were dumped in a local fork of the river that it was given the odiously evocative name Bubbly Creek: a foul stretch where methane and hydrogen sulphide bubbles of

---

Illinois's northern boundary was extended nearly 50 miles (80 km) north so that the planned canal would fall entirely within the same state.

decomposition could be seen at the surface.[41] A far cry from its distinctive shade of turquoise today – except on St Patrick's Day, when it is dyed emerald green – the river was so sludgy that people could walk across it, and so oily that it was prone to setting on fire.[42] The Chicago River, the natural connection between the lake and the lands to the west that had once been so attractive to Indigenous and white traders and settlers alike, was barely a river any more.

To deal with what the novelist Upton Sinclair described in *The Jungle* as a 'great open sewer', another adjustment to Chicago's environment was sought.[43] Following a similar but ultimately flawed attempt to deepen the Illinois and Michigan Canal (the easternmost 11 km/7 miles of which have since been covered by the Stevenson Expressway Interstate 55), a new replacement canal was dug across the ridge dividing the Mississippi River and Great Lakes drainage systems, deep enough that, once connected with the Chicago River, water would flow by gravity not into Lake Michigan, but towards St Louis. The plan was a triumph. Opening in 1900, the Chicago Sanitary and Ship Canal effectively reversed the Chicago River so that sewage would head far away from the lake.[44] To Chicago's city leaders, victory had been achieved over nature, and earth's rules of physical geography had been overcome. Decision-makers in Washington DC, scheming to construct an even bigger canal in Central America, felt vindicated in their faith that humans can conquer any challenge the planet throws their way; later, several of the Chicago canal's engineers brought their techniques and expertise to Panama.[45] And, in a final flourish, when St Louis, now downstream of Chicago's wastewater, sued Chicago in what was the first pollution case to be tried in the US Supreme Court, it lost.[46] (Chicago was not so lucky with injunctions filed by lakeside states concerned that their water supply was being drained away, necessitating the construction of a harbour lock to control water flow into the redirected river.)[47]

Contrasting starkly with the naturally meandering North

Branch, the Chicago Sanitary and Ship Canal can be seen on a modern-day map of Chicago shooting straight as an arrow through the city's south-west. Enabling the movement of boats and sewage alike, it remains symbolic of our ability – and, for some, tendency – to modify the world around us. Rather than adapting to the natural environment and mindfully preserving it, as generations of Indigenous people had done before, history took a sharp turn that saw newcomers shaping their environment around themselves, rather than the other way around. Chicago's ridges tell an analogous story.

Running as elongated lines of relatively high ground, ridges had long been used by Indigenous groups to travel great distances across the wetlands. Where these routes intersected, different communities would meet and trade: a natural geographic advantage that the first white settlers quickly acknowledged, seeking to erect their cabins close to as many busy routes as possible.[48] And so, in the same way that newcomers learned from Indigenous communities how best to travel by water, so too did they come to understand the value of ridges in traversing the region's marshes and finding appropriate places to set up base. Chicago's special convenience as both a water junction between Lake Michigan and the Chicago River, and a ridge junction for numerous popular Midwestern trails, made it irresistible to newcomers. Having gravitated here, some settled permanently in the first years of the nineteenth century, as was true of both John Kinzie and Jean La Lime, fur traders from Quebec (who happened to be at the centre of the first of many murder mysteries in the city).[49] Others, feeling suitably familiar with the region's ridge-based geography, eventually opted to establish new trading posts further along one of the ridges, as was the case of one of Kinzie's Quebecois employees, Antoine Ouilmette, at Gross Point (today Wilmette) to the north. Regardless of what they chose, the Chicago River's confluence at Wolf Point quickly became the primary hub, where Indigenous and white traders readily exchanged business and booze.[50]

While Indigenous people were being incrementally forced from their lands – which actually undermined many of the earliest white settlers, too, for the region's fur trade was decimated in the process[51] – a new group of white settlers from the East Coast had begun flocking to the region in growing numbers from the 1810s. Conscious that the ridges connected some of the most fertile land imaginable, and safe in the knowledge that people had been successfully cultivating the prairies here for generations, farmers quickly established great ranches on plots previously cleared by Indigenous groups, which they could farm faster and faster from 1837 thanks to the development of a polished plough by John Deere.[52] Once they were ready to go to market, all farmers then had to do was follow the routes used for generations by Indigenous people to meet at the main crossroads – Chicago.

There remained one major problem, however: these narrow paths were not fit to accommodate heavy traffic. Following autumn rains or spring snowmelt, the trails would turn to thick mud, causing horses to sink as if the ground were quicksand, and compelling passengers of top-of-the-range stagecoaches to help push their rides out of the mud. In winter, ruts would freeze, making travel bone-jarringly bumpy.[53] Travelling in summer was often optimal, but even then, baking heat and dusty tracks created just as much discomfort. Some stagecoach companies preferred to travel along Lake Michigan's beaches instead, so severe was the strain and time commitment required to reach Chicago via the trails.[54] As early as the late 1830s, attention turned towards the possibility of building railways, these being more capable of traversing the Midwest's mucky terrain through the night as well as the day, better suited to carrying large loads more cheaply and less vulnerable to the region's climate extremes.[55] Thanks to these early railways, Chicago's continued primacy within the Midwest became pretty much inevitable, for not only could they combine with Lake Michigan's ships to offer farmers and merchants lower transport costs and superior economies of scale, but also the city's ideal position between the main coastal termini of New York and San Francisco gave it a significant natural geographic advantage over its main lakeside rival to the north, Milwaukee. By the start of the twentieth century, Chicago boasted more lines of track in more directions and connecting a larger number of individual railway depots than any other North American city.[56] Both physically and conceptually, it was both the link and the boundary between east and west.[57] Still, business in Chicago was certain to be limited as long as roads remained rudimentary, necessitating a search for all manner of solutions.

The Green Bay Trail, which today comprises portions of Rush Street and Clark Street before heading northwards towards Wisconsin, was one such example. Here, white settlers attempted to straighten the naturally crooked path each winter by running a sled through the snow to mark a new route.[58] Further south, a government-sponsored attempt in the early 1830s to widen and straighten the trail that today forms part of State Street – one of the city's main shopping streets – was quickly abandoned due to the challenges posed by building on mud and marsh. The project was only renewed in the late 1860s thanks to the vision and deep pockets of the department store magnate Potter Palmer.[59] In the western suburbs, meanwhile, some ridges were built up higher above the marshes, but here too the new roads were doomed to failure, becoming unmanageably viscous in wet weather. Only in the late 1840s did new plank roads constructed by private corporations represent a reasonable improvement: though prone to rotting and warping, they proved far more capable of withstanding the city's rapidly increasing traffic until gravel and brick roads took their place.[60] After many varying attempts, it seemed that the natural landscape would be tamed – no matter the cost.

Offering an early blueprint for urban development, it is revealing that even after a large swathe of Chicago burnt down in 1871 and a new city plan was conceived, former Indigenous trails were retained and renovated rather than being concealed in a completely consistent, monolithic grid.[61] For instance, whereas the enormous West Market Hall on its route was demolished following the Great Fire, the historic Lake Street Trail continued to be revamped in order to help farmers access a new meatpacking, wholesaler and market centre in the West Loop, a district whose enduring foodie tradition is now manifested in its numerous Michelin-starred restaurants.

It is also no coincidence that many of the first horse-drawn omnibus and horse-car routes from the 1850s followed previously converted Indigenous trails.[62] That said, of all the long segments to survive to the present day, the most famous is surely the old Southwest Plank Road, which follows the trail used by the Potawatomi on their eviction in 1835. Now mostly called Ogden Avenue (named for the city's first mayor, a canal and railroad executive whose attorney was a young Abraham Lincoln), in the early twentieth century the road became part of Illinois's first paved road, the Pontiac Trail between Chicago and St Louis, and was soon after designated as the first section of the country's most celebrated highway.[63] Winding through the lands of some two dozen Indigenous nations all the way from Chicago to Los Angeles, Route 66 may no longer exist officially on US road maps, but as a symbol of connectivity, imprinted on landscapes and minds alike, its significance to all manner of Americans is without peer.

It is incontrovertible that Chicago's Indigenous trails have been crucial to the city's development and its continued status as a major transport hub. And yet, despite influencing settlers and planners' location and logistics decisions, they live largely outside general public consciousness today. Rather than being recognised as formative, the bones of the new city's layout, they have progressively been papered over. Conventional wisdom that Chicago dates to the early nineteenth century risks treating its setting as a *tabula rasa*, on which a world city freely sprouted from paludal environs. The reality, as we know, was rather more contentious and arduous.

An unanticipated declaration of outside interest in the region's Indigenous geography came in 1900–1, with the publication of a map that remains unrivalled in depicting the societies of a century earlier. Through travelling widely across the region, the German amateur cartographer and archaeologist Albert F. Scharf gathered an impressive body of information that is essential to our understanding today, for many of the places he visited were either erased or hidden by new developments in the years thereafter.[64] Among his map's most eye-catching elements are the Des Plaines River running north to south and the gently snaking trails radiating across the plains, all presented in intricate detail. By contrast, even though the modern-day downtown appears in a prominent position, few features are presented other than Fort Dearborn and the natural mouth of the Chicago River, plus the trails that unite in its general vicinity.[65] Also explicitly marked are Indigenous villages and mounds, most of which sit along the region's waterways and ridges, evincing these communities' familiarity with the natural advantages offered by what at first might appear to be a monolithically flat and swampy landscape. The main node for the southern trails, Blue Island, which new settlers promptly converted into a vibrant way station between Chicago and the former capital of Indiana, Vincennes, is given particular prominence. Named for its enigmatic appearance from a distance, sitting on a glacial bluff above the low mist,[66] Blue Island exemplifies Indigenous communities' skill in identifying the most opportune locations for travelling, meeting and trading – which in turn offered settlers a directory for the best places to establish a home and business among the wetlands.

Scharf's map is unlikely to be found on mugs or T-shirts in Chicago's countless gift shops. These and other repositories of the city's collective consciousness instead perpetuate different kinds of geographical connection, associated with Chicago's identity as a modern, international city: its rapid transit lines, five of which convene as an idiosyncratic elevated loop downtown;

American architect Daniel Burnham's 1909 urban plan to recreate 'Paris on the Prairie', with broad boulevards connecting about a dozen parks along a single line. With double- and triple-decker streets insulating grandiose avenues from unseemly freight transport, and a lakefront trail of green spaces and beaches running nearly unbroken for 30 km (18½ miles), the 'City in a Garden'* is a place where order and beauty, form and function exist in their own kind of connective harmony. Scharf's map – and even his decision to produce a map of a hidden feature of the city's landscape in the first place – might imply that Chicago's historic trails are incongruent with such a vision. Bending unpredictably across wetlands now built over, and meeting at villages whose offshoots are now dismissed by some urbanites as generic pieces of the city's sprawl, the trails on the map articulate an alternative system of geographical connectivity, in which earth shaping is primarily dictated by the natural environment rather than by human resolve. Such a conception is at clear odds with how most Chicagoans view their city today, but that doesn't mean that it's outdated or wrong – quite the opposite. In excavating the city's skeleton, buried beneath tarmac and steel, the map allows us to appreciate a different sort of interconnective landscape, of natural ridges and waterways substantially refashioned, but not entirely lost.

More so than perhaps any other city on earth, Chicago is a synthesis of many of the other motives we have seen for doggedly building a connective geography. Akin to the Qhapaq Ñan centuries earlier, albeit with a closer eye on nature than on rival societies, Chicago's street configuration and engineering accomplishments demonstrate a very human desire to bring order to a messy world. Like the Panama Canal, whose own emergence owed somewhat to Chicago's blueprint, the city's potential as a convenient nexus has long been appreciated by logistics professionals and vendors

---

* The Latin phrase *urbs in horto* has been Chicago's motto since 1837.

alike (although maybe not by those changing flights at today's overcrowded O'Hare). As a city built on lands seized from their stewards so that others could extract their natural bounties, Chicago shares certain commonalities with Mozambique's story, while in its determination to reimagine how people internationally work, live and play, it might be considered a modern precursor to Saudi Arabia's postmodern THE LINE. Chicago is unmistakable proof of our ability to reshape our planet in accordance with our cravings and demands: it is a marvel of engineering, a paragon of humans' ability to surmount the resistance nature poses when we try to fashion more conveniently connective landscapes. However, what is easily forgotten among these philosophies of harnessing and wielding geography is that earth shaping can also assume alternative, more nuanced forms – the types Chicago's original inhabitants have lived by for far longer.

Not unlike how the Great Green Wall demonstrates a concern with restoring debilitated lands across borders, Chicago's Indigenous communities, in all their diversity, underscore the need to preserve and fortify those connections that already exist, not least the ridges and waterways that naturally crisscross the Midwest. Indigenous communities here recognised and still recognise the natural interconnections between everything: that Chicago comprises wet*lands*, where land cannot be discriminated from water, and where humans must live in balance with nature's gifts.[67] Instead of filling in the wetlands to create dry land that can be sold and developed, or replacing naturally flowing rivers with rigid canals to ease commerce, all in the name of deriving as much value from the land as possible, respect how land and water *together* provide life. Instead of building a metropolis in a place that still consistently floods after heavy rainfall, or intensively farming the prairies to breaking point, or introducing invasive species that can ravage a region's woodlands and endanger endemic plants and animals, respect what nature offers us, rather than moulding and remoulding it to our hearts' desire. The inspiration for conserving the environment – only belatedly a priority for most of us – was always there. We were just too distracted by shiny new toys to notice.

The case of Chicago should remind us that earth shaping isn't singular. The colonial settler belief that land is to be divided, owned and exploited – 'Let us conquer space,' as future US vice-president John C. Calhoun famously stated in 1816 – is just one way of working with and writing ourselves onto the landscape. A traditional Indigenous understanding that stresses careful management of the connections between land and water, and between living organisms and inanimate entities, is another. No doubt many of us sit somewhere in between, appreciating the privileges that modern infrastructure provides while trying to reduce our impact on our natural surroundings, and hoping that we leave the planet to future generations in a better state than when we ourselves arrived into the world. The issue is that although earth shaping can be manifested in multiple forms, it is seldom fair or equitable. As a result, it is the narratives of the powerful that typically appear most prominently on our planet's palimpsests, while their rivals (and especially those that have been present for longer) are obscured in the process.

Take Chicago. The city's Indigenous history hasn't been completely obliterated from its contemporary landscape. Like many other US cities, its name is derived from an Indigenous language, most likely the Miami-Illinois word *shikaakwa*, meaning 'wild garlic' or 'striped skunk', and possibly both these smelly entities.[68] Peeling away the layers of the Big Onion, one can identify Indigenous names in local streets (Wabash, Sangamon, Wabansia, Winnemac) and villages (Skokie, Winnetka, Waukegan, Algonquin) across the metropolitan area. And zooming out, Illinois is one of a majority (albeit a narrow one) of US states whose name originates in an Indigenous language[69] – neighbouring Indiana's does not, despite its pretence.

Inevitably, plenty of other cities have Indigenous stories

hidden in their own urban palimpsests to tell as well. Arguably New York's two most famous streets, Broadway and Wall Street, were respectively once a Wickquasgeck trail and a Dutch barrier against Indigenous and English adversaries. Looking much further afield in Australia, Sydney's meandering Oxford Street follows an old Indigenous track along a ridgeline in the Eastern Suburbs.

However, in a city that loves to erect honorary signs celebrating its sports heroes, public servants, neighbourhood business owners and more, it is remarkable how few of Chicago's streets, official or honorary, acknowledge Indigenous contributions to the city's ultimate development. More brazen still, in lieu of the connections identified and cherished by Indigenous communities for centuries, Chicago has, with characteristic fastidiousness, concealed its former trails and their histories by naming the roads that supplanted them after white people and places. Even though hundreds of thousands of people technically follow in the footsteps of their Indigenous forerunners every day, few realise that their routes once connected not the various locations characterising a commercial and industrial metropolis, but an array of itinerant communities who chose periodically to congregate at this point between Lake Michigan and the Des Plaines River.

Certainly, hidden behind the contemporary city's idiosyncratic facade of structure and experimentation – two qualities that evince Chicago's unyielding commitment to modernity – merely traces of the former Indigenous trails remain. Within the entire metropolitan area, only the Sauk Trail explicitly recognises an Indigenous group,[70] and because it weaves nearly 50 km (30 miles) south of the downtown area, never entering the city's boundaries, it is little known to many in the city proper. By contrast, a former trail bisecting the Loop is named, apparently to little controversy, after a military officer notorious for his harsh treatment of Indigenous people: George Rogers Clark, the brother of William Clark. To twist the knife further, towards the city's northern limits, Clark Street intersects another street whose own sordid history has been obscured over time. This road marks part of the boundary lines established by the 1816 Treaty of St Louis to prohibit Indigenous people from living near the Chicago River – a detail that was hushed up in 1909, when it was renamed from Indian Boundary Road to Rogers Avenue, in tribute to a local Irish-born settler and landowner.[71] Taken together with plaques honouring the same treaty rather than the people it marginalised, and various examples of Indigenous people being represented in ways both reductive (for instance, in the decontextualised name Indian Boundary Park) and offensive (the same park's fieldhouse and playground depict Indigenous people in feathered headdresses, an attire reflective of white stereotypes of all Native Americans rather than being traditional to the Potawatomi who once lived here),* this failure to commemorate the city's Indigenous history is an act of symbolic violence.

After all, plaques and names have great power: they provide explicit links to the past and subtly yet unambiguously convey to future generations the people and values a society wants to champion. Curiously, not least given the cunningly enshrouded Indigenous origins of many of Chicago's thoroughfares, street names tend to be regarded as quite banal and innocuous relative to more tangible commemorative features of the urban land-scape, even where the figure or event they represent is the same. Christopher Columbus, inflictor of unprecedented misfortune on America's Indigenous populations, is a case in point. Chicago removed its Columbus statues in the wake of protests and van-dalism following the racially motivated police murders of George

---

* A better-known case of misrepresentation concerns the city's National Hockey League team, the Blackhawks, whose assertions of honouring the great Sauk warrior are frequently countered by the view that plastering cartoonish profiles of his face on jerseys and merchandise perpetuates harmful stereotypes of Indigenous people, and moreover constitutes cultural appropriation by a largely white organisation.

Floyd and Breonna Taylor in 2020, yet so far it has retained the name Columbus Drive for one of its most famous downtown streets.* While it's true that a statue or monument provides an overt target of resistance owing to its sheer tangibility (as the United States has also seen in recent years with respect to Confederate memorials), it is quite illuminating that this and other street names have seldom been subjected to the same intensity of conversation concerning the appropriateness of the figures they represent.

It is distressing enough that innumerable white figures with a history of quashing Indigenous people are honoured in street names throughout the city – one that was built directly on top of the trails and crossroads dear to Indigenous traders. Aggravating matters is the general lack of attention given to Chicago's first permanent settler, a man who personally embodied respect and appreciation for Indigenous people. Now, one may well argue that Jean-Baptiste Point DuSable is recognised perfectly well here. A double-decker bascule bridge,† a bust and a museum of Black history all bear his name, and in 2021 the city dedicated Lake Shore Drive to DuSable as well,[72] even if for the meantime, most residents continue to default to this road's old moniker. However, when one considers that the fascist Italian aviator Italo Balbo, whose relevance to Chicago is limited to the fact he led an impressive squadron of seaplanes to the city's 1933–4 world's fair, has been honoured with both a street and a monument for

---

* Still, the situation is fluid. In March 2024, one alderman proposed renaming the street after former US president Barack Obama, given both his direct relevance locally and the city's general paucity of tributes to Black members of the community; a previous petition sought to do similarly for DuSable. Many people and public institutions are similarly polarised on whether Chicago should replace the federal holiday Columbus Day with 'Indigenous Peoples' Day'.

† Fun fact: with thirty-seven, Chicago claims to have the second highest number of operable moveable bridges in the world, after Amsterdam.

the best part of a century, the important legacy of Jean-Baptiste and his wife Kitihawa seems oddly understated; she is still not permanently honoured in the city at all. Yet here was a part-immigrant, part-Indigenous couple who worked hard to manage both a business and a family, who could speak several languages between them and who were crucial to forging trusting relations with everybody, no matter their skin colour or ancestry. In an alternative universe, the DuSables might have been held up as fundamentally American heroes, representatives of an inclusive and equitable United States at around the same time as the country won its independence. In their stead, Chicago chooses to celebrate the itinerant Frenchmen Marquette and Jolliet for their role in exploring and 'discovering' the area, while ignoring the essential support they received from Indigenous communities in navigating its wetlands. A metal sculpture accentuates 'their' revelation at the west end of the old portage, a bas-relief sculpture depicting their passage through the Chicago River is embedded in the very bridge dedicated to DuSable,* and the lobby of an early skyscraper bearing Marquette's name is decorated with some rather dubious mosaics of Indigenous people designed by Tiffany & Co. Together with a suburban city in the case of Jolliet and a city park in the case of Marquette, these men's names are fairly conspicuous in Chicagoland. One must scratch far deeper to unearth a detailed picture of the city's origins.

Today, it is hard to imagine Chicago as anything other than a hub, an indisputable centre of connectivity and influence with the power to shape not merely the Midwest, but the entire world. During rail's golden age, it was the United States' train capital. At one point in the jet age, it was home to the world's busiest

---

* Reflecting a degree of change in how Chicago's history is perceived, previously there had been proposals to name this bridge after either Marquette and Joliet or Fort Dearborn instead.

passenger airport. Now, it is the host of the planet's biggest agriculture commodity exchange and its third largest intermodal port. Here is a city that for nearly two centuries has welcomed people with big American dreams hoping to prosper in almost any endeavour imaginable, and share their concepts and products with others from far away. From continental time zones to Wrigley's chewing gum, McDonald's Hamburger University to the phrase 'getting laid',* Chicago's contributions to contemporary society are as inescapable as they are eclectic. Yet none of this would have been possible without the ingenuity of Indigenous communities. It was they who shared their knowledge of the Chicago Portage, which later made the Illinois and Michigan Canal possible. The canal in turn helped feed the city by conveying agricultural produce to the young metropolis, whose rapid growth catalysed the need for faster and more efficient forms of transportation, the railway and later the aeroplane. In addition, the Indigenous trails offered a handy guide to suitable trading spots among the marshy terrain, which would eventually evolve into Chicago's downtown Loop and many of its earliest suburbs. Chicago may be widely regarded as an exemplar of a modern city – of incredible feats of engineering, of networks and flows, of rampant consumption and unprecedented scale – but it would never have become this without the pioneering work of the Potawatomi, Ojibwe, Odawa and other Indigenous communities that knew not only how to withstand the region's manifold geographical challenges, but also how to benefit from them without injuring nature in the process.

Having long been marginalised not merely by the original exodus but by a questionable federal strategy to relocate and assimilate Indigenous people in major US cities in the mid-twentieth century as well,[73] only recently have Indigenous people's contributions to the region's history and their engagement with

---

* Supposedly this originated as a contraction of 'getting Everleighed' at the infamous but apparently luxurious brothel the Everleigh Club.

its geography started to gain some belated recognition. Since 2021, the world-famous natural history Field Museum has provided dedicated spaces where Indigenous voices can be heard and listened to, an overdue but necessary step in the right direction for an organisation historically accused of hoarding and misrepresenting stolen Indigenous artefacts.[74] Though far less renowned, a second art initiative called 4000N explicitly seeks to promote the power of connection by creating an interactive trail and healing space linking two newly created earthen mounds on the Des Plaines and Chicago rivers together. By encouraging individuals to walk from one end to the other, this 'interpretive learning experience' hopes to re-energise Indigenous people's intimate relationship with the natural world, while educating others about a gentle form of earth shaping that does not demand that local environments be comprehensively modified in order to matter.[75]

Still, no affirmation of the city's Indigenous roots was more explicit and more deliberate than the Ojibwe and Chicago-based artist Andrea Carlson's enormous mural on Chicago's panoramic Riverwalk from 2021 to 2024. For in declaring 'YOU ARE ON POTAWATOMI LAND,' not only did it overtly acknowledge one of the region's most noteworthy Indigenous communities, but quite subtly, it also challenged the precise terms of the treaties signed two centuries ago to displace Indigenous communities. Specifically, the land it sat on did not exist until decades later, when debris from the Great Fire of 1871 was recycled to expand the lakefront (including contemporary Streeterville, Millennium Park and Grant Park, all just east of the city's most iconic road, Michigan Avenue) as part of another undisguised earth-shaping operation. Arguing that they had never ceded this land via any treaty, in 1914 the Pokagon Band of Potawatomi Indians took Chicago all the way to the US Supreme Court, asserting that they had a valid claim to what is now among the most economically valuable land in the city. Predictably, they were ruled as having abandoned the land in 1833, even though (a) it did not exist at the time, and (b) Chicago's Potawatomi community had been coerced to leave the region rather than choosing to do so.[76] Though the mural was incapable of reversing Chicago's history or reverting its synthetic environments to their natural state, in providing an unmissable proclamation that this land was once used very differently, it compelled viewers to remember that the reason this lakeside location even hosts a settlement is because Indigenous people had previously demonstrated a way.

It is indisputable that Chicago's landscape continues to manifest a connective worldview and historical narrative of a society quite different from its Indigenous past. From its monuments to white explorers (not just Columbus, Marquette and Joliet, but Leif Erikson as well),* to its skyscrapers housing many of the firms one would expect in a global city, the metropolitan area quite effectively conceals the fact that with 65,000 inhabitants, it is today home to the largest urban Indigenous population in the Midwest and the second biggest east of the Mississippi River.[77] As its constantly changing riverfront testifies, the city still tends to prioritise bold technological adjustments, yet slowly, alternative conceptions of how we should engage with the planet are, if not quite coming to the fore, at least gaining some of the recognition they deserve. Particularly formidable is an Anishinaabe-led[†] initiative called Mother Earth Water Walk, whereby participants hike the entire perimeter of the Great Lakes to raise awareness of

---

* Meanwhile, the best-known and most public representations of Indigenous people in the city only exoticise their subjects in accordance with a distinctly settler perspective, presenting a bowman and a spearman without clothes. Ironically, one of Chicago's few public examples of Indigenous culture is not even from the Midwest: a totem pole comprising a thunderbird and a whale tail, *Kwanu'sila* in Lincoln Park was created by the Indigenous sculptor Tony Hunt, but is actually traditional to the Kwakwaka'wakw of British Columbia.

† The Anishinaabe comprise several culturally related groups including the Potawatomi, Ojibwe and Odawa.

human damage to these essential, interconnected water bodies, the largest fresh surface-water system in the world. Due to events such as this, more and more people are coming to view the natural world not separately from human society, as a resource to be exploited or an obstacle to be fixed, but as part of the basic fabric of all our lives, requiring solicitous forms of earth shaping that allow both to be revitalised conjointly. The Midwest's Indigenous communities continue to offer us ways of living with nature and appreciating the geographical connections it offers rather than co-opting them to our exclusive benefit. The question, as ever, is will enough of us take notice?

8

Vitality: The Baekdu-daegan

Our country is the three-thousand-ri* golden tapestry-like land with the same mountain range.

Kim Jong-il[1]

---

* A traditional land measure in Korea.

Scale:
0 — 100 Miles
0 — 200 Kilometres

Gando

Cheon-ji
Baekdusan

China

Manchuria

North Korea

Madaesan

Sea
of
Japan

Hwangaksan
Pyongyang

Geumgangsan
Demilitarized Zone (DMZ)

Seoraksan

Kaesong

Odaesan

Bugaksan

Dutasan
Cheong-oksan

Seoul

Hangang

Taebaeksan

Yellow
Sea

Woraksan

Sobaeksan

Songnisan

South Korea

Hwangaksan

Deogyusan

Jirisan

East
China
Sea

Japan

International
border
Baekdu-daegan
Secondary ridges
River
City
Mountain

Jeju-do
Hallasan

Home of probably the world's most infamous border, the Korean peninsula might not seem like an obvious place to search for shared affinities and connections. After all, having been divided officially in 1948 with the creation of two rival states that would quickly be regarded as exemplars of the nascent Cold War era – a democratic South versus a communist North – at first glance, the Korean twins could hardly have grown up more differently. Although long associated with the manufacture of steel, automobiles and electronics, enabling *chaebols* (family-owned conglomerates) such as Samsung, Hyundai and LG to become global household names, South Korea's exports today are increasingly of a cultural variety, from BTS to bibimbap and 'glass skin' to *Parasite*. Doubtless millions will instantly recognise the horse-and-lasso choreography of 'Gangnam Style', or the teal-green tracksuits of *Squid Game*. Perhaps in hopes of better appreciating their favourite K-pop band or K-drama, or immersing themselves on a holiday to Seoul, by the end of 2020 more users of the language app Duolingo were learning Korean than the far more widely spoken Mandarin, Russian or Portuguese.[2] It would be fair to assume that most aren't putting in the hours to prepare for a casual jaunt to North Korea's capital, Pyongyang. Despite offering a welcome of sorts to limited numbers of Chinese tourists, the North, unlike its southern sibling, remains notoriously resistant to establishing a wider tourism industry, and those who have managed to visit typically report that any sightseeing is carefully controlled by the country's leadership. Meanwhile, its cultural production remains almost inconceivably insular, heavily curated and oriented towards a leadership whose distinctive penchant for pomp and circumstance is enough to deter most onlookers abroad.

And so, while the South rides the *Hallyu* wave of its own trendiness, allowing it to define what is considered cool far beyond its borders – except of course in North Korea, where women must refer to male partners as 'comrades' instead of *oppa*,* where Supreme Leader Kim Jong-un regards South Korean pop as a 'vicious cancer',[3] and where skinny jeans, nose and lip piercings and dyed hair are banned – the North's influence and power are expressed in the form of pervasive surveillance, water-tight border control and alarmingly impressive military capabilities. Seventy years after the squiggly Demilitarized Zone (DMZ) replaced the horizontal thirty-eighth parallel north as the boundary between the competing states, authorisation to cross the Korean border is practically unheard of: even when Donald Trump took twenty steps into North Korea in June 2019 as part of a characteristic-ally unorthodox diplomatic gambit, he appeared to rely on the supreme leader's verbal consent rather than any sort of official paperwork.[4] Efforts to cross the DMZ surreptitiously, most often from North to South but on occasion in the other direction, have typically provoked gunshots, in many cases fatal.

From lives to livelihoods and siblings to security, much was lost when Korea, which had been just one state for almost a mil-lennium, was severed by foreign powers following the Second World War. Although division was intended only as a temporary administrative measure while Soviet and US forces expelled the Japanese army, rapidly cooling relations between the two rising superpowers, exacerbated by the young United Nations' failure to find a solution for reunification, ultimately resulted in the estab-lishment of two separate countries in 1948. The new leaders,

---

* An affectionate honorific traditionally used by female Koreans to address an older brother, but which in South Korea is now used for male partners and slightly older male friends as well.

Kim Il-sung in the North and Syngman Rhee in the South, shared perhaps two traits in common, neither of which was conducive to mutual agreement and understanding: principled inflexibility, and a proclivity for autocracy.[5]

Whereas the South sought to renounce the traditional concept of *sadae* ('serving the great', a form of bilateralism placing Korea as the secondary party) and to position itself as a passionate devotee of the capitalist and anti-communist cause, the North committed itself to an extreme version of Marxism–Leninism which it now calls *juche* ('self-reliance').[6] Even before well-organised and -equipped North Korean troops launched a full-scale invasion of the South in 1950, thousands had been killed either as militants or as civilians caught in the crossfire of various border skirmishes.[7] The Korean War, which history books say ended with an armistice in 1953 despite the fact a peace treaty has still never been signed, claimed over 4 million more lives, at least half of whom were civilians, and exacerbated the tragic phenomenon of dispersed families dating back to the oppressive period of Japanese colonial rule (1910–45).[8] Following partition and warfare, approximately three-quarters of a million families are believed to be divided,[9] and although some have managed to reunite in the diaspora – a UNESCO-recognised special live broadcast in 1983 accounts for several thousand[10] – many millions remain unable to access necessary channels of engagement. And with continued tensions between North and South, the DMZ, the heavily fortified, 4-km (2½-mile)-wide border buffer zone slicing Korea in two since 1953, shows few signs of disappearing any time soon.

Koreans globally are united in the loss and trauma of the mid-twentieth century, which prove that more than mere vestiges of the Cold War (and before) endure to the present day. Less recognised internationally, however, are the ways in which the border between North and South not only prevents the movement of people, but also, it is widely believed, weakens the mystical energies shaping the peninsula's geography. The crux of the matter is a continuous line of 487 mountains and hills, dotted with sacred structures and running a good 1,400 km (900 miles) from Baekdusan* on the Chinese border to the Cheonwangbong peak of Jirisan in the Southern province of South Gyeongsang (although some extend the line conceptually from there to Hallasan on the idyllic Jeju-do, thereby connecting Korea's two tallest mountains).[11] This mountain range, named the Baekdu-daegan (literally 'White-Head Great-Ridge'), acts as the Korean peninsula's backbone in a physical sense, bisecting it not west to east like the DMZ, but north to south. Yet it also operates as a backbone in a spiritual sense, as a conduit of Korea's power which must be harnessed and augmented, rendering it central to Korean cultural identity as a single national community, regardless of what the modern-day DMZ may imply.

The belief that the Baekdu-daegan has divine significance has its roots in geomancy, the practice of interpreting various features of the planet and connecting them with people's fortunes.[12] Inspired by Chinese *feng shui* (literally 'wind-water'), Korean *pungsu-jiri-seol* ('wind-water-earth principles theory') is said to have been established by the Seon (Zen) Buddhist monk Doseon-guksa[†] in the ninth century.[13] Having studied sundry esoteric Daoist and Buddhist teachings, Doseon travelled widely across the Korean

---

* A quick note on the language used in this chapter: in Korean, the suffix *san* means 'mountain', *do* means 'island', *guksa* means 'master', *bong* means 'peak', *gang* means 'river', *gung* means 'palace', *sanmaek* means 'mountain range', *ji* means 'lake', *gu* refers to a type of district and *dong* a kind of neighbourhood. Also, I have opted for the English transliteration Baekdusan, to be consistent with Baekdu-daegan, but a reader should know that some sources instead spell this mountain 'Paektusan'.

[†] Doseon remains a mysterious figure even in Korea: the majority of information on his life is derived from legends passed down many years after his death, and so the extent of his role in shaping Korean geomancy is hard to reliably pin down.

peninsula, paying particularly close attention to its mountains, rivers and lakes, for he believed that an auspicious place is one that both stores favourable wind (such as gentle summer breezes, as opposed to far more ominous typhoons in summer and Siberian gales in winter) and provides water.[14] Bringing his theoretical knowledge and convictions together with his personal observations, Doseon's conclusion was both elegant and logical: as an upland area where each of Korea's major rivers originate, the Baekdu-daegan range feeds and conveys spiritual and material vital energies, *gi*,[*] to rivers, fields and people throughout the peninsula.[15] As long as this *gi* flow is unimpeded, the health and prosperity of the Korean nation, Doseon claimed, was guaranteed.[16] In this respect, he succeeded in adapting some of the principles of *feng shui* – whose focus tends to be on burial sites and interior spaces such as the home[17] – to Korea's broader, exterior geography.

Further, Doseon suggested that humans have a role to play in enhancing *gi*. Most notably, he recommended that holy sites be constructed strategically on *hyeol*,[†] sensory openings on the (generally southern) slopes of every mountain, to help balance the earth and heaven's positive and negative energies,[‡] which converge here. Particularly attractive locations are mountains that offer three *hyeol* on plateaus at different altitudes, connected by *gi* lines running down the southern slope: typically, large monasteries are built on the lower level where *gi* is most balanced, while hermitages are constructed on the central and upper *hyeol*, where their

---

[*] The equivalent of *qi* in the Chinese tradition.

[†] Alternatively called *myeong-dang*.

[‡] This notion of opposite but necessarily interconnected forces is consistent with the much older Chinese principle of yin–yang. In the Korean context, this is most clearly represented in the *taegeuk* symbol at the centre of South Korea's national flag, but more conceptually, one may also note that according to traditional Korean geomancy, mountain ranges are regarded as yin while streams are considered yang.

residents can access more refined *gi*.[18] Ideally, too, three mountains and hills form a horseshoe providing protection from cold northerly winds, but open to the south; in such cases, each mountain has a designated mythological creature, with the tallest 'black tortoise' standing to the north, a 'white tiger' to the west and an 'azure dragon' to the east, plus a fourth 'vermilion bird' further south.* Doseon also championed Korea's traditional form of shamanic practice, Mugyo or Muism, which is characterised by the worship of *san-shin* mountain spirits, to increase the flow of *gi*.[19] (Reflecting later geopolitical tensions between Korea and China, Korean legend claims that one such spirit had originally introduced Doseon to examples of auspicious sites at Jirisan, which nationalists have sometimes used to buttress their arguments that *pungsu-jiri* is distinctly Korean as opposed to being largely rooted in Chinese *feng shui*.)[20]

Doseon believed that through supplementing the earth's desirable natural energies in these ways, people are simultaneously capable of curing the land and improving the wisdom and health of those living nearby, a principle he called *bibo*, or harmony with nature.[21] In this regard, the careful management of *hyeol* essentially represents a geographical version of acupuncture, the traditional Chinese medical practice of improving energy flows in the human body by inserting needles in specific places.[22] In fact, even today some liken the Baekdu-daegan to the human body, its mountains collectively forming the spine and spinal cord, consolidating the peninsula and carrying energy in the manner of the central nervous system.[23] The range also bears certain similarities to the cardiovascular system, and it is noteworthy that *hyeol* means not only

---

* Readers familiar with Chinese philosophy (*Wuxing*), mythology or astrology may recognise these animals as the 'Four Symbols' or 'Four Guardians' (*Sì Xiàng*), each of which represents an element as well as a cardinal direction. Hence the black tortoise represents water, the white tiger metal, the azure dragon wood and the vermilion bird fire.

'hole' as a geographical descriptor, but more conceptually, 'blood' as well. As such, the most sacred mountain, Baekdusan, acts as a heart, from which the peninsula's lifeblood is pumped initially through the Baekdu-daegan and then through secondary ridges, rivers and streams until it reaches either a settlement or the sea.[24]

Doseon died in 898 CE, but his ideas were fundamental to the previously warring territory's unification over the next few decades. On ascending to the throne in 918, Wang Geon (also known as Taejo of Goryeo) ordered the construction of numerous sacred structures at *hyeol* across the peninsula, and designated Gaegyeong, now Kaesong, as his new capital, based on the monk's advice. While it presumably didn't hurt that Gaegyeong also happened to be Wang Geon's hometown and stronghold, these actions were important for two key reasons. First, they represented something of a declaration that *punsu-jiri* should play a prominent role in the new dynasty; and second, they indicated that the state would enjoy significant jurisdiction over how geomantic knowledge would be used moving forward.[25] Wang Geon's death in 943 would only reinforce the state's tight relationship with *pungsu-jiri*, thanks to an injunction he had dictated stating that his successors would be permitted to build Buddhist temples exclusively in those places selected by Doseon, lest the dynasty fall upon new hardships.[26]

From this point on, geomancers duly came to play a critical role in the evolution of Korean society, tasked with monitoring *gi* flows to key settlements and identifying alternative sites whenever they seemed beyond repair.[27] Many held prominent positions in the government – geomancy was even included in the Goryeo Dynasty's civil service exam from 958 – and were not afraid of ruffling a few feathers.[28] A striking example was a respected Buddhist monk and political adviser to the dynasty named Myocheong, who in the 1130s seized Seogyeong, the previous name for Pyongyang, and declared it the capital of a brand-new state. Why go to such efforts? Citing serious religious and political misgivings concerning Goryeo's future, he was frustrated at the royal court's refusal to move the seat of government from Gaegyeong, whose energy forces he claimed were waning, to this more northerly ancient capital. Myocheong's suggestion was grounded on more than a little self-interest: Seogyeong was his hometown, and he appeared determined to install his own loyalists from this city in the government. However, his combined geomantic and geostrategic expertise enabled him to argue, quite persuasively,* that making Seogyeong the capital would be in the nation's best interests, it being both far more auspicious according to *pungsu-jiri* and more convenient for a future assault on the Confucian Jin Dynasty of north-eastern China. The city would much later fall upon great misfortune, including sieges, battles and the bombastic whims of megalomaniacs, but in the first instance, it was Myocheong instead who was out of luck: his rebellion was crushed and he was put to death.[29]

Stories such as these epitomise how Korea's human geography is a palimpsest of the nation's commitment to *pungsu-jiri*. Where an otherwise attractive site was deemed to be lacking in one minor detail, humans would carefully alter the landscape, for instance by erecting an artificial hill to supplement or better conserve locally weak energies,† or by building a pagoda or shrine or simply renaming local places, as symbolic compensation for a topographic deficiency.[30] Similarly, rural communities might achieve harmony with nature by remoulding complex or flawed local landscapes into the shape of a symbolic animal or object, with the effect of personifying them and rendering them more familiar and meaningful.[31] Through its cities, temples, tombs

---

* The king agreed to build a new palace in the spot recommended by Myocheong, but stopped short of moving the capital, infuriating the monk and provoking his rebellion.
† On a much smaller scale, some real-estate developers today develop artificial mountains in city neighbourhoods in order to boost local buildings' auspiciousness and market value.

and more, we can uncover why certain places were developed and when, why some cities were seemingly abandoned, and why others were revived when the time was right.[32] At the everyday level, court cases have been won and lost over geomancy, while burial sites have long been selected based not on the preferences of the now-deceased, but on whether their descendants might prosper in future.[33] We might also consider areas that don't appear to have ever undergone much change, or whose transformations were especially subtle. Since a decline in *gi* was believed to portend loss of fortune, land or safety, the Joseon Dynasty from the late fourteenth century opted to implement a policy of prohibiting human activities that might 'cut' or damage mountain ranges near settlements to ensure the continuous flow of the Earth's energy.[34]

Moreover, although some dismiss it as mere superstition, one should consider how *pungsu-jiri* additionally champions tangible geographical advantages. For example, Doseon's pinpointing of *hyeol* recognises the ability of mountains and hills to tame wind and offer sunny slopes, water sources and protection from floodwaters (not to mention potential invaders). Sacred *maeul-soop* groves can improve water conservation and shade, act as windbreaks, supply firewood and edible greens to residents and organic matter to the soil, and enhance local biodiversity, as well as counteract natural energy flaws in the landscape.[35] Planting artificial woodlands along streams can mitigate flooding, and doing similarly along newly created ponds both increases local species diversity and provides migrating animals with an interconnected green corridor.[36] The Neo-Confucianist civil servant and geographer Yi Chung-hwan even produced a comprehensive geomantic and geographic survey* of the entire peninsula and, doubtless of great help to home seekers, a guide to the best places to live according to *pungsu-jiri* – an eighteenth-century version of *Niche* or *Business Insider*, if you will.

---

* *T'aengniji* ('Ecological Guide to Korea').

Regardless of whether their planners and architects complied with *pungsu-jiri* for primarily spiritual or practical reasons, it is no coincidence that significant sacred sites including temples and pagodas are most commonly found on south-facing slopes and rarely on north-facing ones.[37] In some older villages, every single building is oriented in one direction, a curious quirk of Korean human geography that owes to a specific kind of *hyeol* championed by Doseon called a *baesan imsu* ('mountain in back, river in front').[38] These low-altitude locations, where a field or plain cut by a river provides prosperous *gi* to the front, while a mountain to the rear (connected of course to the Baekdu-daegan) calms the wind, are considered felicitous for human habitation. The best-known example of a settlement that was developed on a *baesan imsu* is Seoul, Korea's (though now just South Korea's) capital for all but a six-year interregnum since it replaced Kaesong in 1394. Traditionally, new dynasties would seek to increase their fortune and political legitimacy by moving the seat of government, and so, on ascending to the throne in 1392, Yi Seong-gye (like Wang Geon, alternatively named Taejo) of the nascent Joseon Dynasty consulted prominent geomancers to identify the most auspicious site possible.* Hanyang, as Seoul was called back then, was practically peerless: behind it to the north is the mountain Bugaksan, and in front of it, facing south, is a stream, the Cheonggyecheon.†

---

* The Joseon Dynasty interpreted *pungsu-jiri* slightly differently from its Goryeo predecessors, largely on account of its Neo-Confucian rather than Buddhist foundations, although the fact that it retained the key philosophical principles is testament to the system's mainstream approval by this time.
† The modern life of the Cheonggyecheon has been rather more eventful than the king had probably anticipated in the 1390s. Having become a repulsive makeshift sewer after the Korean War (1950–3), it was covered up first with concrete and second by an elevated highway, transforming it into a geographical connection of a different kind. Don't expect to see this highway today, however: it was dismantled from 2003 as part of a pricy urban renewal and beautification project that has allowed the stream to be restored and

With wind further stored by two protective rings comprising eight total mountains around the city – four of which are specifically regarded as guardians protecting Seoul from catastrophes* – and extra water provided by the larger Hangang, it is hard to imagine a more geomantically attractive place to live.[39]

With its new capital in place, during its 500-year lifetime the rulers of the Joseon Dynasty worked hard to benefit further from the earth's natural energies, carefully constructing five grand royal palaces in shrewd locations between water and mountain in the districts today called Jongno-gu and Jung-gu, and filling in a gorge to improve the city's connection with the mountains to the north. They also acted to suppress the dangerous fire energy they identified emanating towards their city from Gwanaksan south of the river by erecting two sculptures of *haechi* (called *xiezhi* in China) at the main gate of the fourteenth-century Gyeongbok-gung, the city's largest palace and the best positioned according to *pungsu-jiri*.[40] These mythical unicorn-lions were thought capable of providing protection from blazes; whatever one believes today, they remain beloved among the city's residents, were made Seoul's official emblem in 2009 and now appear, often alongside cartoon versions of the city's four mountain guardians, in parks, outside subway stations and on buses, taxis, T-shirts and food packaging.

Another (albeit subtler) indicator of Seoul's geomantic significance is Myeong-dong, a bustling neighbourhood known best today for its retailers. In Korean, *myeong* means 'bright', which is appropriate enough considering how the area's illuminated signs are an assault on the senses, but in the shamanic-Daoist tradition, this word is used to imply that a place is *sacred*.[41] Keeping this

---

made available to the public once again. No longer despised, it is one of the contemporary city's chief leisure and tourism assets, a linear park furnished with distinct wetlands, miscellaneous art installations and a museum.

* Along with the northern Bugaksan, these are Naksan in the east, Namsan in the south and Inwangsan in the west.

information in mind, it maybe becomes a little bit less remarkable – but no less interesting – why so many popular names for
people and places insinuate brightness, including Kim, 'gold'. The
fourth most common Korean family name, Choi, has a spiritual
relevance of its own: it means 'mountain'.[42]

Given this general commitment to the principles of *pungsu-jiri*,
it makes sense that the prominence of the Baekdu-daegan within
the Korean psyche has only been reinforced over time, too. In the
early Joseon period, Baekdusan in the far north was regularly
privileged as the peninsula's principal energy source, although the
Baekdu-daegan as a fully integrative concept had yet to emerge.[43]
The oldest existing Korean world map, the Gangnido (1402),*
is a case in point. Whereas the peninsula's shores are generally
ill-defined, the Baekdu-daegan is carefully and unmistakably
depicted as a dark line running parallel to the east coast, with
smaller ridges and rivers branching away to the west. However, the
map suggests a discontinuity around a quarter of the way south
from Baekdusan, reflecting the fact that the Baekdu-daegan was
permeated with deeper, connective significance to the Korean
people later, in the eighteenth century, amid a burgeoning of
national consciousness during a border dispute with the Chinese
Qing Dynasty.[44] A particularly noteworthy proponent of this
ideology was the renowned Neo-Confucianist scholar 'Seongho'
Yi Ik (1681–1763), who, in describing the Baekdu-daegan as
'the repository of great figures', drew popular attention to the
mountain range's historical as well as geomantic significance to
Korea.[45] Most prominent among these Koreans was the first ruler
of the enduring Joseon Dynasty, Yi Seong-gye, whose familial links
to Baekdusan perfectly complemented this mountain's status as
what Yi Ik considered the 'ancestor' of the Baekdu-daegan, and
helped justify the decision to officially designate this mountain

---

* Officially known as the *Honil Gangni Yeokdae Gukdo Ji Do* ('Map of
Integrated Lands and Regions of Historical Countries and Capitals').

as the birthplace of the whole kingdom in the 1760s.[46] Reflecting this fusion of human relationships and the natural environment, Neo-Confucianists still often describe the Baekdu-daegan's peaks as a family, with Baekdusan representing the great patriarch and its southern analogue Jirisan being the matriarch.[47]

Quickly consolidating this rather emotive notion of upland-oriented national identity was a second publication, an unprecedentedly systematic study of Korea's mountains called *Sangyeongpyo*. As dry as this 'Table of Korean Mountains' may sound, its importance lay in the fact that it explicitly articulated the Baekdu-daegan as a tangible connection throughout the national territory. Specifically, having assumed the unenviable task of categorising every Korean mountain by location and size, the geographer Shin Gyeongjun (1712–1781) described how the main mountain system (the *daegan*) also comprises fourteen second-ary mountain ridges (known as *jeonggan* and *jeongmaek*), which direct the peninsula's greatest rivers through separate watersheds towards the sea.[48] According to the notion that 'mountain ranges divide streams' – a principle that endures to this day – a person can thus feasibly walk from one end of the Baekdu-daegan to the other without ever crossing a body of water.[49] Shin's methodical approach was very unconventional at the time in Korea, for tra-ditionally geographical analysis here meant seeking and sensing *hyeol* intuitively rather than collecting objective, empirical data. Even so, he retained a somewhat familiar approach by explicitly connecting the government offices of different towns to local mountains (hence the boundaries of any district or region could be defined by its physical geography), consistent with the older and enduring Goryeo convention of linking administrative boundaries with mountain ranges and rivers.[50] And so, with the production of increasingly accurate maps and growing awareness of how one's settlement or region was defined by the Baekdu-daegan, the distinctly sensory Korean worldview became more formalised, teaching that not only is the natural world characterised by total connectivity (both among mountains and between mountains and rivers), but it exists in complete intersection with the human world, too.

Exemplifying the esteem in which it is held, the Baekdu-daegan brings together sacred sites with all four of Korea's major abiding religious and philosophical traditions, including Buddhism, Daoism, Muism and Neo-Confucianism; Protestantism, though conspicuous in its popularity today, remained marginal here until the nineteenth century.[51] Artistic depictions of Korea have long emphasised themes of connectivity, flow and the paramount status of Baekdusan both literally and symbolically. For example, the Baekdu-daegan has been painted as the spine and stripes of a tiger (traditionally Korea's national animal, symbolising power and protection, although North Korea now prefers the mythological winged horse *chollima*), and as the trunk of a tree, embellished by either branches or roots for the smaller mountain ridges. Further, while these fourteen mountain ridges are sometimes regarded as boundaries between Korea's cultural regions, including in terms of cuisine, fashion and dialect,[52] the fact that they are all attached simultaneously allows Koreans to see themselves as connected, as *one*. The Baekdu-daegan, then, is far more than just a mountain range: it is a potent cultural symbol of Korean cultural identity and nationhood.

Over the past 300 or so years, which have seen Korea's nearest geographical rivals grow and grow, this conceptual power has continued to swell, providing the basis for an independent Korean identity in response to threats from abroad. A perfect example dates from 1712, when the Qing Dynasty commissioned the placement of a boundary pillar to clarify its frontier with the Korean Joseon Dynasty. It quickly transpired that this stone had been inscribed with a description of the border that placed a portion of Korean territory, including Baekdusan (which is called Changbaishan in China), on the Chinese side of the line. This act, which inadvertently drew much of the Korean public's attention to the sacred mountain and provoked new, jingoistic feelings that it was key to the nation's territory, did not simply smack of rank opportunism: tracing their genealogy back to the mythical hero Bukūri Yongšon, whom they believed had been conceived near Baekdusan when his mother ate a piece of magical fruit, the Qing leadership also wanted to claim ownership of what they considered their ancestral lands.*

Fearful of Qing pressure both real and symbolic along the border, in the 1760s King Yeongjo inaugurated a new tradition of praying and offering rites to the Baekdusan deity, in hopes of emphasising this borderland's importance from a Korean perspective. However, the boundary-stone controversy re-emerged in the 1870s and 1880s, when, having been threatened with expulsion by the Qing court, Koreans living in the border Gando region (Jiandao in Chinese) forcefully pointed out that once again, China was intentionally misleading the public as to the precise location of the border. Representatives from both sides were assigned the perilous task of ascending Baekdusan to identify which of two local rivers actually marked the frontier.[53] The Korean perspective won out on this occasion – a clear point of pride for the nation's cartographers, who consistently chose to present the boundary stone on their maps – although again the dispute did not end there.[54] Some in China accused Korea, incorrectly, of moving the stone further north so that Baekdusan would be included within

---

* Like the earlier Jin Dynasty, the Qing Dynasty originated in Manchuria, China's north-easternmost region to the immediate north of Korea, and regarded much of the area around Baekdusan as sacred. A separate but relevant point on regional reverence: in the Qing foundation myth, the magical fruit is dropped by a magpie. This detail helps explain not merely why these birds were considered divine in Manchuria, but also why many people in China still believe they bring good luck, starkly at odds with the traditional British superstition that lone magpies are harbingers of misfortune.

its territory,[55] an allegation that became moot when imperial Japan ceded Gando and the entirety of the sacred mountain to China in 1909. Despite this redrawing of Korea and China's borders, the monument actually remained in place until it was mysteriously removed in 1931, seemingly in preparation for a more clearly deceitful border controversy: the Japanese army's detonation of a small bomb on a Japanese-owned railway near Mukden (today Shenyang), as pretext for a full invasion of north-eastern China.[56] North Korea would only reclaim the southern side of the mountain in the 1960s as part of a secret border agreement with China, and eagerly placed new boundary markers to celebrate the achievement.[57]

More broadly, the rise of Japan as a modern imperial power posed a grave threat to Korea and its distinctive relationship with its mountains. Whereas Japanese forces had been repelled by the allied Joseon and Ming dynasties three centuries earlier, around the turn of the twentieth century Japan tightened its stranglehold on the East Asian mainland, assassinating Empress Myeongseong (also known as Queen Min) in 1895, before making Korea a protectorate (1905) and then a colony (1910).[58] Over the next thirty-five years, Japan reconceptualised Korean geography according to its own land-surveying preferences, which meant dismissing the notion of a unitary Baekdu-daegan mountain system and replacing it with the more international concept of individual mountain ranges, with the effect that the shorter 'Sobaek-san-maek' was conceptually cleaved from the 'Taebaek-sanmaek'.[59] No less egregiously from a Korean perspective, it thoroughly modified Korea's once-familiar landscapes, dispossessing farmers of their land, clearing large parcels of forest, and constructing new settlements, railway lines, roads, dams and mines throughout its sacred mountain range.[60]

Korea was thus rapidly transformed from a primarily agrarian society into an industrial colony oriented to serving Japan's needs, a different form of connectivity that consolidated the latter's control both practically and symbolically. And while Korea's soils and rivers were exploited mercilessly in the interests of industrialisation, over time its cultural distinctiveness was jeopardised by a forced assimilation policy: students were forced to speak Japanese and worship at Shinto shrines, Korea's shrines and monuments were destroyed, and countless places, including Korea itself,* were renamed.[61] By the end of the Second World War, more than three-quarters of a million Koreans had been displaced to Japan and its colonies to serve as forced labourers,[62] but even among those who stayed, Korea quickly became unrecognisable.

The legacy of Japanese colonialism remains controversial in Korea from a geomantic standpoint. In the 1980s–90s, a peculiarly fierce debate centred on a widespread accusation that Japan had committed a kind of *feng shui* terrorism on Korea, cunningly driving iron stakes into its mountains in an effort to sever *gi* flows and thereby subdue Koreans' patriotism and vigour.[63] Although it's plausible that the posts were actually used for land surveying rather than anything more sinister – and South Korea's rapid economic growth since the early 1960s would surely refute the claim that the country is destined for misfortune as long as a few metal bars are stuck in the ground – the tale is occasionally still recycled. As a different kind of needle-like structure, wind turbines have long represented a particular bugbear,† especially in regions whose significant elderly populations have long memories.[64]

A more recent episode illustrates how much conceptual-geographical trauma continues to exist from the colonial period, even in the highest office of the land. Since South Korea's establishment in 1948, Cheong Wa Dae (the 'Blue House') has served as the executive office and residence of the country's president, but

---

* Chōsen.
† Another argument against wind turbines here is that they make noises and flash lights that apparently resemble those of *dokkaebi*, fire-wielding goblins in shamanic folklore, which must be gratified lest they bring harmful energy.

in 2022, the newly elected leader, Yoon Suk-yeol, opted instead to move into the presumptive foreign minister's official residence in a neighbourhood further south, closer to the river. Although the precise reasons for Yoon's decision remain a mystery, it has not escaped the attention of *pungsu-jiri* devotees that Korea's Japanese occupiers had built an official residence for the governor-general on this site as part of a larger strategy to deprive the beloved Gyeongbokgung of *gi*. The Japanese had already constructed their colossal General Government Building directly in front of the fourteenth-century palace, which, just for good measure, they were also in the process of systematically demolishing. The new governor-general's residence effectively completed this act of geomantic and symbolic hostility, sandwiching Gyeongbokgung between two overt symbols of Japanese power while severing *gi* flows from both the northern and southern mountains.[65]

Independence allowed the abhorred General Government Building to be destroyed instead, while Gyeongbokgung continues to undergo lengthy reconstruction and restoration work, but still many see the Cheong Wa Dae site as deeply unlucky[66] – and not without justification. Among the presidents to have lived there, Park Chung-hee was assassinated, Choi Kyu-hah was forced to resign following a coup, Roh Moo-hyun was impeached and later died by suicide, Kim Young-sam's presidency was marred by the 1997 Asian financial crisis, and Chun Doo-hwan, Roh Tae-woo, Park Geun-hye and Lee Myung-bak have all served prison terms. It is certainly not out of the question that Yoon wanted to improve his fortunes by converting Cheong Wa Dae into a public park, while basing himself in Hannam-dong, which is located propitiously by the Hangang, backed by the guardian mountain Namsan. However, unfortunately for Yoon, this move has not exactly had the desired effect: following a botched attempt to impose martial law on December 2024, his presidency was brought to an ignominious end.

If anything is depressing Korea's energy flows, it is surely the DMZ. At about 250 km (160 miles) long and splitting Korea in two, it is both a physical barrier preventing the movement of people from one side to the other, and a psychological boundary between two opposing ideologies in one of the world's geopolitical hotspots. Fortified with rows of electric, razor and barbed-wire fences, anti-tank ditches, minefields, explosive charges and high-resolution surveillance cameras and sensor systems, its name belies its status as the most heavily militarised and monitored border in the world. Though hospitable to animals that favour its dearth of people, including rare Asiatic black bears, Amur leopards, Siberian musk deer, cinereous vultures and Manchurian trout,[67] the DMZ exists as a deep wound on the Korean landscape, a source of great pain for families which remains difficult to suture. As long as the northern and southern segments of the Baekdu-daegan remain divided, Korea's future appears bleaker than even the most pessimistic geomancers could have predicted in the decades before.

And yet there is still some reason to believe that reunification will be possible one day, for despite their real and perceived estrangement, the two Koreas continue to share their reverence for the Baekdu-daegan. Partition since 1945 has failed to erode this mountain range's centrality to Koreans' sense of identity, regardless of which side of the border they live. Unsurprisingly, considering its location as well as that the country's founder Kim Il-sung used it as a base for his armed resistance movement against Japan, Baekdusan appears particularly frequently in North Korean symbolism and propaganda. Whereas Soviet sources suggest that Kim Jong-il, North Korea's second supreme leader, was born in Siberia, official North Korean accounts claim that he was born in a secret guerrilla camp on Baekdusan, thereby ensuring that the great leader national myth of *Baekdu hyultong*, the sacred Baekdu 'bloodline', would be sustained.[68] When Kim Jong-il died in 2011, the state news agency KCNA described some bizarre natural phenomena occurring at the mountain and its neighbouring Cheon-ji (Heaven Lake), including a glowing sky, a natural message carved in rock dedicating Baekdusan to his revolution, and ice that cracked so loudly both Heaven and Earth shook.[69] (Sadly, the bulletin did not mention Cheon-ji's version of the Loch Ness Monster, a popular myth on the Chinese Tianchi side of this crater lake, one of the highest of its kind globally.) Additionally, the country's national emblem since 1993 depicts Baekdusan towards its centre, immediately below a communist red star and accompanied by two other primary symbols of the regime, a hydroelectric plant and ears of rice. And whenever a propaganda photo shoot is in order, Kim Jong-un is known to enjoy visiting the mountain atop a white horse, an animal associated nationally with his grandfather, Kim Il-sung. Meanwhile, South Korea's national anthem shares with its northern counterpart a reference to Baekdusan (in its very first line), while it retains the traditional national myth that the mountain is the birthplace of the god-king Dangun, founder of the first Korean kingdom, Gojoseon.[70] Regardless of what the DMZ may suggest about the two present-day countries, Korea remains one people, united by one mountain range.

Catalysed by the June 15th North–South Joint Declaration of 2000, in which Kim Jong-il of the North and Kim Dae-jung of the South agreed to pursue the possibility of cooperation, the past two decades have seen bouts of political traction to improve relations, even if these have yet to translate into durable dialogue over reunification.[71] During his presidency (2017–22), South Korea's Moon Jae-in proffered various gutsy ideas, including a joint bid to co-host the 2030 FIFA men's World Cup, and full reunification by 2045, the centenary of the peninsula's liberation from Japan.[72] So far, humbler initiatives rooted in the Baekdu-daegan – and therefore symbolic of Korean life and connectivity – have proved somewhat more fruitful. For instance, when Moon collaborated with Kim Jong-un in planting a pine tree in the DMZ in 2018 – an act that risked conjuring up uncomfortable memories from 1976, when two US soldiers were bludgeoned to death for pruning a tree that North Koreans believed had been planted by Kim Il-sung – they made sure to use a mixture of soil and water from mountains and rivers on either side of the border.[73] A few months later, the two leaders met again, climbing Baekdusan together before dipping their hands in Cheon-ji. South Korea's first lady, Kim Jung-sook, additionally poured some water she had brought from Hallasan into the lake, bringing the two furthest ends of the Baekdu-daegan together.[74]

The symbolism of this visit was intense. Not only had Moon, a keen mountain climber, managed to check off what might have been the top item on his personal bucket list, but he had also become the first South Korean leader to climb the Korean peninsula's holiest mountain from the North Korean rather than the Chinese side. A casual pastime for many across the world, hiking may in fact be the key to future Korean reconciliation and perhaps even reunification, a feeling that Moon acknowledged:

> Now that the first step has been taken, more will follow in our footsteps, and I believe we are embarking upon an era where average South Korean citizens will be able to climb Mount Baekdu as tourists.[75]

Indeed, since the 1980s the Baekdu-daegan has re-emerged as a modern nationalist symbol for many in South Korea, simultaneously with a growing rejection of colonial-era place names such as Taebaek-sanmaek, and the establishment of a 735-km (450-mile) hiking trail from north to south. Combining environmental, political, nationalistic and spiritual rationales, new conservation laws have been passed over the past twenty years in order to help alleviate the legacy of Japanese colonialism and industrialisation, although these have sometimes proved contentious due to their potential to frustrate local industries, piecemeal administration across different geographical domains, and the difficulties of defining boundaries for a necessarily fluid and abstract rather than explicitly bounded place.[76] Also worth recognising as an earlier stimulus of an environmentally conscious South Korean national identity is the country's mandatory tree-planting effort in the 1960s–80s, which reforested large areas that had suffered from Japanese colonialism and the Korean War.[77] Reflecting Korea's traditional veneration of the mountains, this was a conceptual as well as a merely practical concern, given the belief that all features of the landscape are interconnected and vulnerable places must therefore be healed.

Aided by the publication of detailed maps, magazines, guidebooks and websites in English as well as in Korean, interest in and excitement about the Baekdu-daegan only continues to grow, both domestically and internationally. The size of Scotland and Wales combined, South Korea is today home to twenty-three national parks, eighteen of which are classified as mountains and eight of which are traversed by the Baekdu-daegan ridgeline, including the first and largest, Jirisan. Millions of South Korean city dwellers, clothed in expensive new gear, now choose to escape to the mountains at least once a month to take advantage of the natural beauty and metaphysical substance just a short drive away. Partaking in the modern concept of 'forest bathing' – immersion in nature to rejuvenate mind, body and soul – they reconnect with the mountains historically at the core of what it means to be Korean. Hiking in South Korea is a far more communal activity than tends to be the case elsewhere: strangers will commonly share food and drink at the summit while contemplating their shared connection to the land, an opportunity to build camaraderie and connection while simultaneously reflecting inwardly and growing individually.[78]

North Koreans' modern-day relationship with the mountains is quite different, being focused not on leisure or spirituality, but on more pragmatic matters such as foraging, hunting and collecting firewood. Although a tiny number of lucky foreigners have managed to gain permission to hike with bemused locals in North Korea's near-untouched highlands – an extraordinary opportunity to admire Baekdusan and exchange questions and experiences with the hermit kingdom's residents – crossing the DMZ is currently impossible, and in any case, negligible necessary infrastructure such as trails and shelters exists.[79] At present, the only option available to foreigners seeking to travel to Baekdusan is to join a tour company and fly from Pyongyang by chartered plane. But what if it one day became possible to trek the entire length of the Baekdu-daegan, uninhibited by the DMZ, to appreciate its beauty, its role in Korean history, its palpable sacred power – and maybe even restore those *gi* flows long obstructed and weakened by the DMZ?

This aspiration forms the basis of an ambitious initiative to create a modern-day pilgrimage trail along the Baekdu-daegan crestline. While pilgrimages – typically challenging journeys to sacred sites with the power to transform the participant spiritually – pre-date antiquity, they have not traditionally occurred between Jirisan and Baekdusan, partly on account of the dangers involved in climbing the very steepest slopes, but at least as importantly because of the presence of dangerous Amur or Siberian tigers.[80] These animals were regarded as pests and eradicated in Korea during the Japanese occupation,[81] although given the modern North's geopolitical isolation, it is a mystery whether any can still be found in the far north-eastern corner of the peninsula. However, emboldened by South Koreans' burgeoning love of hiking, as well as more than a little curiosity about the world's most reclusive country, there is incipient hope that the North will one day agree to inaugurate formal routes for walkers, with no intention of upsetting or destabilising its regime, to travel all the way through one of the world's least explored regions to hallowed Baekdusan.[82]

Thus, in the same way that the South is gradually labelling and publishing educational material on both its natural (peaks, passes, rivers) and human-made (monasteries, pagodas, shrines, hermitages, pavilions, altars, charismatic *jangseung* totem poles) points of interest in order that visitors can better appreciate the peninsula's customs and history of religious coexistence,[83] so too might the North publicly list its most significant sites, many of which have been widely forgotten due to neglect. It may even be that the North one day grants the opportunity to experience traditional culture in an equivalent way to that offered by the Templestay programme in the South, which since the 2002 FIFA men's World Cup has permitted visitors to meditate, converse and stay with Seon Buddhist monks and nuns. With the Baekdu-daegan already sharing certain striking similarities to the famed Appalachian range in the United States with respect to its geology and flora, it has the capacity to even exceed it as an iconic trail thanks to its enormous cultural tourism potential, as a place where walkers can savour a magnificent view one minute and explore a fortress or temple the next.[84]

Through such a pilgrimage, traditional forms of spiritual seeking could be updated to suit contemporary expectations, needs and possibilities. It doesn't matter whether a person is primarily motivated by religion, history, adventure or otherwise: everyone, regardless of background, regardless of citizenship, is united by the act of walking along Korea's ridgeline, over its most important rivers, all the way to its holiest mountain. The Baekdu-daegan, for too long a divided mountain range, can become whole again, and the meanings of Koreanness bound up in it can be made real.

For the time being, those with aspirations of hiking the full Baekdu-daegan as a personal act of Korean unity are likely to be left disappointed: travel to North Korea remains harder for South Koreans than for pretty much everyone else,* requiring the

---

* US passport holders have been prohibited by their government from travelling to North Korea since 2017, when an American student named Otto Warmbier, who had been convicted of stealing a propaganda poster, suffered fatal injuries in North Korean custody.

exceptional authorisation of both countries' governments. Still, change is always conceivable. Although people are not currently allowed to enter the country directly from South Korea, for a brief time around the turn of the millennium it was actually possible to cross from South into North Korea in order to visit both the old capital Kaesong and the sacred Geumgangsan. An arm of Hyundai even gained exclusive permission to develop a tourist region near the latter, and hoped eventually to extend tours all the way to Baekdusan. After a South Korean tourist was shot dead by a Northern soldier for wandering into a military zone in 2008, the agreement was promptly terminated by the South; the North has since seized and dismantled what had once been regarded as flagship cross-border infrastructure.[85]

Yet the potential for long-distance recreational walking is surely there: the mountains are too profoundly connected to what it means to be Korean, in every way conceivable, for this notion to be dismissed out of hand. Many individuals in the South now regard a hike along their country's section of the Baekdu-daegan as a gesture of inclusive patriotism, a symbol of a Korean national identity that residents on both sides of the border share.[86] Recognising this new phenomenon, various politicians share images of themselves hiking the crestline as a declaration of their bonds to the Korean people and nation, while some companies have been known to include maps of a fully intact Baekdu-daegan in their adverts.[87] With growing attention to the importance of environmental conservation and preservation, the next step may well be a popular push for a judiciously managed earth-shaping enterprise: the creation of a cross-border biodiversity corridor, capable of engendering ecological benefits for species on both sides.[88] If such a trust exercise were to prove successful, perhaps in time it will become feasible for people to move more freely as well. After all, the opening of the DMZ Peace Trail in 2018 hints at a remote possibility of one day being able to walk further north, as well as inviting introspection of a quite different sort.

Over a millennium after Doseon initiated a distinctively relational conceptualisation of space in Korea, whereby no mountain or hill can be viewed in isolation, the Baekdu-daegan persists as the locus of *Gangsan*, the universally professed 'land of rivers and mountains', a traditional sobriquet for Korea as a whole.[89] In spite of colonialism, civil war and partition, the mountain range's significance lives on; it is a rare unifier in one of the world's most stringently divided regions. Whereas the DMZ is internationally viewed as the most conspicuous geographical feature in Korea, signifying and imposing near-total detachment between North and South, and providing an uncomfortable reminder that a Cold War-style nuclear crisis may not be more than one unsound decision away, millions of Koreans still look to the Baekdu-daegan for a different, more sanguine kind of reminder: that a shared sense of Koreanness still exists. It may be hidden behind fences and explosives, big talk and hostile words, but it has not disappeared. In our desire for peace and harmony in a region that has experienced far more than its fair share of agony, this point may be all we can cling to, but it's something. And were reunification ever to be achieved, as unlikely as this may seem at times, the Baekdu-daegan – the real and symbolic connection between North and South – will doubtless be part of the solution.

Epilogue

How might the world look had our predecessors not perceived, fashioned and reworked the connections they did? Would Spain have remained one of the planet's foremost imperial powers for three centuries had its conquistadors not exploited the Inca roads and looted the Andes' mineral wealth, and would Southern Africa enjoy a more expansive, integrated and dependable railway network had the Portuguese and British truly chosen to cooperate rather than compete? Would trade have remained slower and more localised had the concept of a maritime shortcut in Central America (and, for that matter, Egypt) been abandoned, and would the growth potential of the major economies on either side of the Pacific have stayed limited? Would Saudi Arabia currently be as committed to reimagining urban life without the impetus of global competition, and would Chicago and Seoul, in different eras, have become among the world's largest and most internationally oriented cities in the absence of their respective templates of interconnectivity? Would our modern maps even depict Estonia, Latvia and Lithuania, Panama or Mozambique, had the potential of earth shaping in inspiring new borders or restoring old ones not been actualised? Without the appreciable power of geographical connections, the globe and human society would surely be very unlike our present-day reality.

Imagine, too, if some of history's forsaken connections had in fact been brought to fruition. Consider not merely the Nicaragua Canal and the Cape to Cairo Railway, but also the Grand Contour

Canal, a proposal to boost England's freight transport by linking its principal ports via a waterway whose naturally consistent elevation would obviate the need for regular locks (presumably Ferdinand de Lesseps would have adored this project, had he somehow still been alive in the 1940s). Chicago, a city whose very existence is owed to our obsession with reshaping the earth, might have hosted a tower over three times the height of Dubai's Burj Khalifa today, capable of dispatching riders via toboggan to a selection of cities as distant as Montreal and Boston, had one Mr J. B. McComber's allegedly practicable concept for a monument at the 1893 world's fair not duly been spurned in favour of the relatively pragmatic Ferris wheel.[1] Sharing the same limitless faith in human capabilities that characterised so many Americans of the time, others were rather more confident in the connective advantages of railways, which McComber's plan was partly designed to circumvent:* the former governor of Colorado Territory, William Gilpin, for instance, who proposed that a railway network be constructed tying together all the planet's inhabited continents. According to this scheme, Denver would have become the world's primary rail hub.[2]

It is easy to dismiss such outlandish ideas out of hand, and yet we would do well to remember that sometimes, as the Panama Canal testifies, even the most brazen concepts for newly earth-shaped connections grow legs. As we wait, dumbfounded, to see whether and how THE LINE evolves, we should bear in mind that despite the fact that Gilpin's Cosmopolitan Railway never materialised, his idea of joining Alaska with Siberia via the Bering Strait still might, this being the central pursuit of the InterContinental Railway group, whose ultimate goal is to connect Paris and Istanbul

---

* As McComber submitted in his justification for a 'Toboggan Tower', 'people from a distance can come to Chicago without being called upon by the railroads to furnish a life-size portrait of themselves, as well as furnishing the companies with other information of a delicate nature'.

with New York and, for some reason, El Paso.[3] While most of the planet's modern connective infrastructure projects provide obvious utility in tying together otherwise detached places – the fibre-optic cables that provide internet to hundreds of millions, or Croatia's Pelješac Bridge, which allows road vehicles to bypass two border crossings along Bosnia and Herzegovina's minuscule stretch of coastline – a minority are better known for their unadulterated pizzazz, as is true of Shanghai's Bund Sightseeing Tunnel, whose psychedelic lighting seeks to give tourists the impression of travelling through a hyperactive, illuminated rendition of the earth's crust and mantle. Other initiatives entertain us by reworking pragmatic connections into leisure spaces, such as the various railway lines converted into public parks (the Coulée Verte in Paris and the High Line in New York, to name just two), and Bogotá's weekly routine of reorganising 120 km (75 miles) of its road system into the makeshift bicycle superhighway Ciclovía. As these examples imply, geographical connections do not have to be forged or adjusted for entirely hardheaded purposes. They can also be ideological or idealistic, manifesting the very human ways in which we choose to engage with our planet, from cementing our authority and directing attention towards ourselves, to simply having a bit of fun.

The truth is, forging links is inherent to our nature. Whether we are drawing new desire paths on our daily walks, or choosing to follow in the footsteps of a figure we admire (the Via Dolorosa representing the path trudged by Jesus to his crucifixion springs to mind), we cannot help but write ourselves and our convictions onto the landscape. The connections we produce and reproduce provide a record of our existence, archiving the idiosyncratic ways in which we understand the world around us and encourage others to view it accordingly. What would otherwise be obscure locations have been put on the map, figuratively speaking, owing to the connections fashioned in the past. Individuals such as Pachacuti, Louis Jolliet and Doseon are remembered in large part because of the distinctively interconnective ways in which they viewed their surroundings. And in rarer cases, a figure who might otherwise have been forgotten to all but the most attentive historians is recognised because another person dedicatedly carved their story into the landscape. Over 700 years since the death of Queen Eleanor of Castile, the route followed by her funeral procession is still faintly marked on England's geography by the monuments dedicated in her honour. It scarcely matters that most of these 'Eleanor Crosses' have been lost to time: King Edward I's commemorative act ensured that his beloved wife would be remembered for generations, while transforming a dozen towns and villages into an integrated memorial.

The conceptual power of connectivity is such that our narratives can survive even if the names on the map change. Signs and tourist infrastructure have helped Route 66 live on as a symbol of freedom and adventure, despite the fact that the United States' most iconic road was broken into separate segments of the Interstate Highway System in the second half of the twentieth century. Nor can the landscape bear only one narrative. As much as we may want others to embrace the same stories as us, some hold particular resonance because they are personal or speak to our own identities and experiences: many visitors to Boston follow the Freedom Trail connecting a selection of sixteen historic sites, but others prioritise the Black Heritage Trail or create their own paths, according to the memories they prefer to remember and reproduce. Some of the stories written onto the earth's canvas are unpopular not because they are little known, but because they are associated with a figure whose divisive legacy provokes alternating efforts to commemorate and marginalise them. In contrast to the nostalgia and romanticism of Route 66, the open road par excellence, the Jefferson Davis Highway, a century-old project to designate much of the US highway system as a fully integrated memorial to the only Confederate president, has never inspired the acclaim its architects desired, remaining instead a discernibly incoherent route acknowledged or valued by few.[4] As an evolving document of our actions, voices and memories, the earth's landscapes and their assorted linkages reveal more than merely the routes travelled or the affinities perceived by people in the past. They divulge the secrets of our relationship with the planet and with others, seeking to bolster the places that matter to us, and ensure the stories of those we most admire are retold and appreciated in perpetuity.

Through incrementally inscribing new stories and applying meaning to otherwise ordinary locations, we all contribute to the planetary palimpsest. However, this does not mean that earth shaping is representative or value-free. Just as the Jefferson Davis Highway seeks to present and legitimise an alternative history of the United States, we have seen how THE LINE aims simultaneously to reimagine urban life in general and Saudi society specifically. Not dissimilarly to how Korea's Japanese occupiers were accused of tampering with the peninsula's energy flows in the early 1900s, Chicago's Indigenous communities remain aggrieved at settlers' systematic refashioning of the Midwest's naturally interconnective geographical features in the century before. Earth shaping tends to be a strategic decision, and for this reason, it is generally also political. If we feel that existing connections obstruct or exclude us, we may seek to alter or undermine them, or we may attempt to create entirely new ones, on our terms. At the smallest of scales, humble desire paths, those accidental trails of erosion, manifest our wilful, subconscious disregard for rules and customs, sculpting new routes whose increasingly striking appearance attracts others to join us in challenging an authority's conception of our geography. By marching along freeways following the 2020 murder of George Floyd, Black Lives Matter protesters proved how the connections designed for automobile users could be reworked, albeit temporarily, into protest spaces, whose symbolic intensity extended to the fact that many of the same roads had originally been designed to disenfranchise and detach Black-majority neighbourhoods from the wider urban fabric.[5] (More generally, one might also consider how cities whose transport infrastructure exclusively embraces road vehicles practically and palpably detach non-drivers from a legion of possible destinations.) At a far larger scale are cross-border initiatives such as the gradually expanding multi-modal International North–South Transport Corridor, which allows one of the planet's current pariahs, Russia, to conveniently shift freight to and from its friends India, Iran and Azerbaijan without needing to use more conventional routes such as the Suez Canal. Examining linkages such as these helps us reveal more than just how different actors engage with specific spaces; it allows us to understand how they perceive and seek to improve their 'place' in the world.

Other connections expose a clash of priorities, a truth particularly conspicuous whenever the environment is at stake. Whether our concern is new energy pipelines that risk leaking hydrogen or natural gas, or commercial waterways cutting carelessly through delicate wetlands such as Brazil's Pantanal, there is very good reason to be perturbed about the environmental damage our earth shaping is liable to cause – a reality to which many residents of Panama can testify. Our challenge, then, is to remain diligent and attentive to the connections we are writing onto our landscape, for the benefits of even the most assiduous forms of earth shaping may not be enjoyed for ever. Just as various pre-Inca civilisations disappeared when their connective infrastructure became incapable of supporting them further, the Garamantes of the Sahara collapsed when their ingenious *foggara* groundwater channels finally dried out, and the Romans' growing inability to transport food efficiently from distant farms to their capital, despite their famous roads, was an important factor in their own demise. A paradox of earth shaping, and one with which the Great Green Wall's creators (as well as Mozambique's government and, potentially in future, THE LINE's placemakers) must constantly contend, is that no matter how strong our integrative concept, the connections we create are only ever as resilient as their weakest link. As soon as essential components of the chain are fractured, debilitated or cut off entirely, the entire system is liable to break. The Baltic Way proved that when decisions are made with the unambiguous goal of achieving popular buy-in, a connective vision can flourish. The challenge for ongoing initiatives such as the Great Green Wall is in ensuring that individuals who have long felt marginalised by community projects now see that they, too, can be a part of a paradigmatic, earth-shaping endeavour.

These issues might seem to suggest that our earth shaping is bound to eventually impair the planet and undercut the opportunities available to large numbers of its inhabitants, and yet we know that there is not just one way to engage with our world. As specific communities in Chicago, the Sahel and Korea have long exemplified particularly well, there are always ways of working with rather than against the grain, of using connections as a means of respecting nature, bolstering shared affinities, and in the process realising the precise objective we are trying to achieve. An obvious modern example would be the planet's many wildlife corridors, such as Central America's Mesoamerican Biological Corridor and India and Nepal's Terai Arc Landscape, which enable endangered species to overcome human boundaries and migrate, feed and breed. More subtle are endeavours to emulate a historic version of physical geography: the new footbridge at the legendary Tintagel Castle in England, although clearly constructed by humans, effectively replicates the land bridge weathered and eroded by the sea centuries ago, while greatly reducing erosion by visitors in other parts of the complex. While pilgrimages, despite their semblance of separateness from earthly matters, are not necessarily free of political wrangling, the conscientiousness with which millions of people follow set routes and rules to visit specific sites suggests that geographical connections have a timeless, internationally acknowledged power in strengthening bonds and transforming individuals from within.

Whatever one's faith, or lack thereof, the accomplishment of an integrated Baekdu-daegan trail would surely provide an immediate gold standard for how geographical connections can be used to honour the past and engender optimism for the future simultaneously.

Regardless of whether we are seeking to render our surroundings more ordered, convenient or befitting of our worldview, revitalise or restore beloved landscapes and take advantage of their inherent benefits, resist a rival force, or reimagine how we might engage with the world around us, earth shaping is irresistible. Wherever we might live, and however privileged we might be or consider ourselves, earth shaping is a concern we all hold – one that has moulded the places we know and will continue to guide our planet's future. Our routes and rituals, friendships and practices are intimately tied to generations, even centuries, of fashioning the places we cherish in a certain way. To shape the world around us is inevitable; for the same reason, our planet's geography is not. The earth bears innumerable signs of our tendency to view and engage with the world through connections. The question, always, is whether we can read the clues.

Notes

Introduction

1 Ellie Violet Bramley, 'Desire paths: The illicit trails that defy the urban planners', *Guardian*, 5 October 2018, www.theguardian.com/cities/2018/oct/05/desire-paths-the-illicit-trails-that-defy-the-urban-planners.

2 Carl Ortwin Sauer, *Land and Life: A Selection from the Writings of Carl Ortwin Sauer*, ed. John Leighly (Berkeley, University of California Press, 1963).

3 Notably, for instance, James S. Duncan, *The City as Text: The Politics of Landscape Interpretation in the Kandyan Kingdom* (Cambridge, Cambridge University Press, 1990).

4 Josephine Crawley Quinn, 'A Carthaginian perspective on the altars of the Philaeni', in Josephine Crawley Quinn and Nicholas C. Vella (eds), *The Punic Mediterranean: Identities and Identification from Phoenician Settlement to Roman Rule* (Cambridge, Cambridge University Press, 2014), pp. 169–79.

5 Rose Parfitt, 'Fascism, imperialism and international law: An arch met a motorway and the rest is history …', *Leiden Journal of International Law*, vol. 31, no. 3 (2018), pp. 509–38.

6 Josephine Quinn, 'Libya's ancient borders', *London Review of Books*, 17 March 2014.

7 Parfitt, 'Fascism, imperialism'.

8 Quinn, 'A Carthaginian perspective'.

9 Parfitt, 'Fascism, imperialism'.

10 Alessandro Raffa, 'The *strada litoranea*: Mapping colonial rural landscape along the Libyan coastal road', *SHS Web of Conferences*, vol. 63, no. 06002 (2019).

11 Parfitt, 'Fascism, imperialism'.

12 Sami Zaptia, 'Philaeni statues' disappearance refuted by antiquities authorities', *Libya Herald*, 16 October 2017.

13 Raffa, 'The *strada litoranea*'.

14 Parfitt, 'Fascism, imperialism'.

15 Emilio Distretti, 'The life cycle of the Libyan Coastal Highway: Italian colonialism, coloniality, and the future of reparative justice in the Mediterranean', *Antipode*, vol. 53, no. 6 (2021), pp. 1421–41.

16 Noha Elhennawy, 'Libya's interim government reopens Mediterranean highway', Associated Press, 20 June 2021.

17 Distretti, 'The life cycle'.

1. Order

1 Pedro de Cieza de León, *The Incas of Pedro de Cieza de León*, trans. Harriet de Onís, ed. Victor Wolfgang von Hagen (Norman, University of Oklahoma Press, 1976), p. 138.

2 R. Alan Covey, 'The Inca Empire', in Peter Fibiger Bang, C. A. Bayly and Walter Scheidel (eds), *The Oxford World History of Empire* (Oxford, Oxford University Press, 2021), pp. 692–717.

3 Ruben Martinez Cabrera et al., 'Center of territorial domains and the network of the Inca Road: Qhapaq Nan, in the province of Tumbes, Peru', *3C Tecnología*, vol. 10, no. 4 (2021), pp. 51–87; José Barreiro, 'Introduction: Strong in our hearts', in Ramiro Matos Mendieta and José Barreiro (eds), *The Great Inka Road: Engineering an Empire* (Washington, DC, Smithsonian Books, 2015), pp. 1–10.

4 Charles R. Ortloff, 'Canal builders of pre-Inca Peru', *Scientific American*, vol. 259, no. 6 (1988), pp. 100–7.

5 Michael E. Moseley, 'Chan Chan: Andean alternative of the preindustrial city', *Science*, vol. 187, no. 4173 (1975), pp. 210–25.

6 Bruce Hathaway, 'Endangered site: Chan Chan, Peru', *Smithsonian Magazine*, 2009, www.smithsonianmag.com/travel/endangered-site-chan-chan-peru-51748031.

7 Tierras de los Andes, 'Pikillacta Archaeological Park, a pre-Inca Wari city', 2024. https://terandes.com/en/blog/cusco/archaeological-sites/pikillacta-park.

8 Jean-Pierre Protzen, 'Inca stonemasonry', *Scientific American*, vol. 254, no. 2 (1986), pp. 94–105; William H. Isbell and Margaret Young-Sanchez, 'Wari's Andean legacy', in Susan E. Bergh (ed.), *Wari: Lords of the*

Ancient Andes (New York, Thames & Hudson/Cleveland Museum of Art, 2012), pp. 251–67.

9   Guadalupe Martínez Martínez, 'Qhapaq Ñan: El Camino Inca y las transformaciones territoriales en los Andes peruanos', *Ería*, vols 78–9 (2009), pp. 21–38; Drew Benson, 'Peru resurrects ancient ways of farming', *Los Angeles Times*, 10 August 2003.

10   Luis Guillermo Lumbreras, 'The Qhapac Nan Museums Network', *Museum International*, vol. 56, no. 3 (2004), pp. 111–17.

11   Catherine Julien, 'The Chinchaysuyu road and the definition of an Inca imperial landscape', in Susan E. Alcock, John Bodel and Richard J. A. Talbert (eds), *Highways, Byways, and Road Systems in the Pre-Modern World* (New York, Wiley-Blackwell, 2012), pp. 147–67.

12   John Hyslop, *The Inka Road System* (Orlando, FL, Academic Press, 1984).

13   Ricardo Manuel Espinosa, 'The Great Inca Route: A living experience', *Museum International*, vol. 56, no. 3 (2004), pp. 102–10.

14   Hyslop, *The Inka Road System*.

15   Brian S. Bauer, 'Suspension bridges of the Inca Empire', in William H. Isbell and Helaine Silverman (eds), *Andean Archaeology III: North and South* (New York, Springer, 2006), pp. 468–93.

16   Jeff Brown, 'Highways to empire: The Inca road system', *Civil Engineering*, vol. 86, no. 1 (2016), pp. 40–3.

17   Bauer, 'Suspension bridges'; Lucy C. Salazar, 'Machu Picchu: Mysterious royal estate in the cloud forest', in Richard L. Burger and Lucy C. Salazar (eds), *Machu Picchu: Unveiling the Mystery of the Incas* (New Haven, CT, Yale University Press, 2004), pp. 21–48.

18   Bauer, 'Suspension bridges'.

19   Martínez, 'Qhapaq Ñan'.

20   Mark Cartwright, 'The Inca road system', *World History Encyclopedia*, 8 September 2014, www.worldhistory.org/article/757/the-inca-road-system.

21   Covey, 'The Inca Empire'.

22   George Peter Murdock, 'The organization of Inca society', *Scientific Monthly*, vol. 38, no. 3 (1934), pp. 231–9.

23   Hyslop, *The Inka Road System*.

24   Terence N. D'Altroy, *The Incas* (Malden, MA, Blackwell, 2002).

25   Charles Stanish, 'Nonmarket imperialism in the Prehispanic Americas: The Inka occupation of the Titicaca Basin', *Latin American Antiquity*, vol. 8, no. 3 (1997), pp. 195–216.

26 Espinosa, 'The Great Inca Route'.

27 Martínez, 'Qhapaq Ñan'.

28 D'Altroy, *The Incas*.

29 Francisco Garrido, 'Rethinking imperial infrastructure: A bottom-up perspective on the Inca Road', *Journal of Anthropological Archaeology*, vol. 43 (2016), pp. 94–109.

30 Espinosa, 'The Great Inca Route'.

31 Barreiro, 'Introduction'.

32 Julien, 'The Chinchaysuyu road'.

33 Darryl Wilkinson, 'Infrastructure and inequality: An archaeology of the Inka road through the Amaybamba cloud forests', *Journal of Social Archaeology*, vol. 19, no. 1 (2019), pp. 27–46.

34 Stanish, 'Nonmarket imperialism'.

35 Wilkinson, 'Infrastructure and inequality'.

36 Juan de Betánzos, *Suma y narracion de los Incas*, ed. Marcos Jiménez de la Espada (Madrid, Manuel O. Hernández, 1880), www.cervantesvirtual. com/obra/suma-y-narracion-de-los-incas-que-los-indios-llamaron-capaccuna-que-fueron-senores-de-la-ciudad-del-cuzco-y-de-todo-lo-a-ella-subjeto--0; Jessica Joyce Christie, 'Inka roads, lines, and rock shrines: A discussion of the content of trail markers', *Journal of Anthropological Research*, vol. 64, no. 1 (2008), pp. 41–66.

37 Christie, 'Inka roads'.

38 Dmytro Dubilet, *How the Tricolor Got Its Stripes: And Other Stories about Flags* (London, Profile Editions, 2023).

39 John Edward Staller, 'Dimensions of place: The significance of centers to the development of Andean civilization: An exploration of the *ushnu* concept', in John Edward Staller (ed.), *Pre-Columbian Landscapes of Creation and Origin* (Springer, 2008), pp. 269–313.

40 Staller, 'Dimensions of place'.

41 Catherine J. Allen, '*Ushnus* and interiority', in Frank Meddens et al. (eds), *Inca Sacred Space: Landscape, Site and Symbol in the Andes* (London, Archetype, 2014), pp. 71–8.

42 Ian Farrington, *Cusco: Urbanism and Archaeology in the Inka World* (Gainesville, University of Florida Press, 2013).

43 Staller, 'Dimensions of place'.

44 National Museum of the American Indian, 'Heart of the Inka universe', 2016, https://americanindian.si.edu/inkaroad/inkauniverse/cusco/cusco-experience.html.

45 Mariusz Ziółkowski and Jacek Kościuk, 'Astronomical observations

in the Inca Temple of Coricancha', *Arqueología del Perú*, 11 September 2021, https://arqueologiadelperu.com/astronomical-observations-in-the-inca-temple-of-coricancha.

46 Brian S. Bauer, *The Sacred Landscape of the Inca: The Cusco Ceque System* (Austin, University of Texas Press, 1998).

47 Christie, 'Inka roads'.

48 D'Altroy, *The Incas*.

49 Farrington, *Cusco*.

50 John Hyslop, *Inka Settlement Planning* (Austin, University of Texas Press, 1990).

51 Ernesto Capello, *City at the Center of the World: Space, History, and Modernity in Quito* (Pittsburgh, PA, University of Pittsburgh Press, 2011).

52 'Inca ruins in Ecuador: The Ingapirca archaeological complex', 2020, www.ecuadorhop.com/inca-ruins-ecuador.

53 D'Altroy, *The Incas*.

54 Steven R. Gullberg, *Astronomy of the Inca Empire: Use and Significance of the Sun and the Night Sky* (Cham, Springer, 2020).

55 D'Altroy, *The Incas*.

56 Salazar, 'Machu Picchu'.

57 Patrice Lecoq, *Nouveau regard sur Choqek'iraw (Choque Quirao): Un site Inca au coeur de la Cordillere de Vilcabamba au Perou* (Oxford, Archaeopress, 2014).

58 Cabrera et al., 'Center of territorial domains'.

59 Tom Shearman, 'Incallajta, Cochabamba, Bolivia', 2018, www.andeantrails.co.uk/blog/incallajta-cochabamba-bolivia.

60 Charles Stanish, *Ancient Titicaca: The Evolution of Complex Society in Southern Peru and Northern Bolivia* (Berkeley, University of California Press, 2003).

61 Carolyn Dean, *A Culture of Stone: Inka Perspectives on Rock* (Durham, NC, Duke University Press, 2010).

62 Christie, 'Inka roads'.

63 Christian Vitry, 'Mountains and the sacred landscape of the Inka', in Ramiro Matos Mendieta and José Barreiro (eds), *The Great Inka Road: Engineering an Empire* (Washington, DC, Smithsonian Books, 2015), pp. 34–8.

64 Christie, 'Inka roads'.

65 Dean, *A Culture of Stone*.

66 Covey, 'The Inca Empire'.

67 R. Alan Covey, 'Kinship and the performance of Inca despotic and infrastructural power', in Clifford Ando and Seth Richardson (eds), *Ancient States and Infrastructural Power: Europe, Asia, and America* (Philadelphia, University of Pennsylvania Press, 2017), pp. 218–42.

68 Christie, 'Inka roads'.

69 Dean, *A Culture of Stone*.

70 Brown, 'Highways to empire'.

71 Maxim Samson, *Invisible Lines: Boundaries and Belts That Define the World* (London, Profile Books, 2023).

72 John Pemberton, *Conquistadors: Searching for El Dorado: The Terrifying Spanish Conquest of the Aztec and Inca Empires* (London, Futura, 2011).

73 D'Altroy, *The Incas*.

74 Patricia Seed, 'Exploration and conquest', in Thomas H. Holloway (ed.), *A Companion to Latin American History* (Oxford, Wiley-Blackwell, 2011), pp. 73–88.

75 Ibid.

76 D'Altroy, *The Incas*.

77 Michael J. Schreffler, *Cuzco: Incas, Spaniards, and the Making of a Colonial City* (New Haven, CT, Yale University Press, 2020).

78 Lumbreras, 'The Qhapaq Nan'.

79 Marie Arana, *Silver, Sword, and Stone: Three Crucibles in the Latin American Story* (New York, Simon & Schuster, 2019).

80 Kris Lane, *Potosí: The Silver City That Changed the World* (Oakland, University of California Press, 2019); Christian Volpe Martincus, Jerónimo Carballo and Ana Cusolito, 'Roads, exports and employment: Evidence from a developing country', *Journal of Development Economics*, vol. 125 (2017), pp. 21–39.

81 José Antonio Camuñez and Kennedy Rolando Lomas, 'The Inca Trail (*Qhapac Ñan*) as a contribution to sustainable tourism in Ecuador', in Andrea Basantes-Andrade, Miguel Naranjo-Toro, Marcelo Zambrano Vizuete and Miguel Botto-Tobar (eds), *Technology, Sustainability and Educational Innovation (TSIE)* (Cham, Springer, 2020), pp. 406–20.

82 Martínez, 'Qhapaq Ñan'.

83 Covey, 'The Inca Empire'.

84 Miguel Angel Carpio and María Eugenia Guerrero, 'Did the colonial mita cause a population collapse? What current surnames reveal in Peru', *Journal of Economic History*, vol. 81, no. 4 (2021), pp. 1015–51.

85 Camuñez and Lomas, 'The Inca Trail'.

86 Patrick Symmes, 'What it's like to travel the Inca Road today', *Smithsonian Magazine*, 2015, www.smithsonianmag.com/innovation/what-its-like-travel-inca-road-today-180955740.

87 Hugh Thompson, *The White Rock: An Exploration of the Inca Heartland* (Woodstock, NY, Overlook Press, 2003).

88 Symmes, 'What it's like'; Bauer, 'Suspension bridges'.

89 D'Altroy, *The Incas*.

90 Annick Benavides, 'The treasure of Coricancha', *Ñawpa Pacha*, vol. 43, no. 2 (2023), pp. 139–73.

91 C. Cuadra et al., 'Preliminary evaluation of the seismic vulnerability of the Inca's Coricancha temple complex in Cusco', *WIT Transactions on the Built Environment*, vol. 83 (2005), pp. 245–53.

92 Bill Sillar, 'Landscape biography of a powerful place: Raqchi, Department of Cuzco, Peru', in Justin Jennings and Edward R. Swenson (eds), *Powerful Places in the Ancient Andes* (Albuquerque, University of New Mexico Press, 2018), pp. 129–74.

93 David L. Bowman, 'New light on Andean Tiwanaku', *American Scientist*, vol. 69, no. 4 (1981), pp. 408–19; 'Laja', www.lapazlife.com/laja.

94 Christie, 'Inka roads'; National Museum of the American Indian, 'Resistance and adaptation', 2016, https://americanindian.si.edu/inkaroad/invasion/resistance-adaptation.html.

95 R. Alan Covey and Kylie E. Quave, 'The economic transformation of the Inca heartland (Cuzco, Peru) in the late sixteenth century', *Comparative Studies in Society and History*, vol. 59, no. 2 (2017), pp. 277–309.

96 Daniel W. Gade and Mario Escobar, 'Village settlement and the colonial legacy in southern Peru', *Geographical Review*, vol. 72, no. 4 (1982), pp. 430–49.

97 Jeremy Ravi Mumford, *Vertical Empire: The General Resettlement of Indians in the Colonial Andes* (Durham, NC, Duke University Press, 2012); Martínez, 'Qhapaq Ñan'.

98 UNESCO, *Qhapaq Ñan Andean Road System: New Steps Towards Its Sustainable Conservation* (Paris, UNESCO, 2021).

99 Wilkinson, 'Infrastructure and inequality'; Hannah Sistrunk, 'Road to empire: Documenting an Inca road in northern Ecuador', *Ñawpa Pacha*, vol. 30, no. 2 (2010), pp. 189–208.

100 Espinosa, 'The Great Inca Route'.

101 UNESCO, *Qhapaq Ñan*.

102 Christie, 'Inka roads'.

103 'Ayni: Ceremony to the Pachamama', 23 May 2017, www.enigmaperu.
com/blog/ayni-ceremony-to-the-pachamama.

104 Hyslop, *The Inka Road System*.

105 Ana Paula Franco, Sebastian Galiani and Pablo Lavado, *Long-Term Effects of the Inca Road*, National Bureau of Economic Research (NBER) working paper 28979, 2021.

106 Sonia Alconini, *Southeast Inka Frontiers: Boundaries and Interactions* (Gainesville, University Press of Florida, 2016), p. 15.

2. Extraction

1 Quoted in Charles Mohr, 'The men who will control Mozambique: Their aim is more than "black power"', *New York Times*, 21 September 1974.

2 Portuguese Historical Museum, 'Mozambique', https://
portuguesemuseum.org/?page_id=1808&category=3&exhibit=&event
=142.

3 Hugo Silveira Pereira, 'Colonial railways of Mozambique: Critical and vulnerable infrastructure, 1880s–1930s', *HoST: Journal of History of Science and Technology*, vol. 16, no. 1 (2022), pp. 7–28.

4 Maria Paula Diogo and Bruno J. Navarro, 'Re-designing Africa: Railways and globalization in the era of the new imperialism', in David Pretel and Lino Camprubí (eds), *Technology and Globalisation: Networks of Experts in World History* (Cham, Palgrave Macmillan, 2018), pp. 105–28.

5 Pereira, 'Colonial railways of Mozambique'.

6 Chris Alden, *Mozambique and the Construction of the New African State: From Negotiations to Nation Building* (Basingstoke, Palgrave, 2001).

7 Ana Carneiro et al., 'Portuguese engineering and the colonial project in the nineteenth-century', *International Committee for the History of Technology*, vol. 6 (2000), pp. 160–75.

8 Diogo and Navarro, 'Re-designing Africa'.

9 Alan K. Smith, 'The idea of Mozambique and its enemies, *c.*1890–1930', *Journal of Southern African Studies*, vol. 17, no. 3 (1991), pp. 496–524.

10 Diogo and Navarro, 'Re-designing Africa'.

11 Smith, 'The idea of Mozambique'.

12 Diogo and Navarro, 'Re-designing Africa'.

13 Smith, 'The idea of Mozambique'.

14 Joshua Kirshner and Idalina Baptista, 'Corridors as empty signifiers:

The entanglement of Mozambique's colonial past and present in its development corridors', *Planning Perspectives*, vol. 38, no. 6, pp. 1163–84.

15 Hugo Silveira Pereira, 'The camera and the railway: Framing the Portuguese empire and technological landscapes in Angola and Mozambique, 1880s–1910s', *Technology and Culture*, vol. 64, no. 3 (2023), pp. 737–59.

16 Pereira, 'Colonial railways of Mozambique'.

17 Hugo Silveira Pereira, 'Colonial railways and conflict resolution between Portugal and the United Kingdom in Africa (*c.*1880–early 1900s)', *HoST: Journal of History of Science and Technology*, vol. 12 (2018), pp. 75–105.

18 Sören Scholvin and Johannes Plagemann, 'Transport infrastructure in Central and Northern Mozambique: The impact of foreign investment on national development and regional integration', South African Institute of International Affairs, 2014.

19 Pereira, 'Colonial railways and conflict resolution'.

20 Diogo and Navarro, 'Re-designing Africa'.

21 Robert Williams, 'The Cape to Cairo Railway', *Journal of the Royal African Society*, vol. 20, no. 80 (1921), pp. 241–58.

22 Pereira, 'Colonial railways of Mozambique'.

23 Pereira, 'The camera and the railway'.

24 Pereira, 'Colonial railways and conflict resolution'.

25 Diogo and Navarro, 'Re-designing Africa'.

26 Smith, 'The idea of Mozambique'.

27 Pereira, 'Colonial railways of Mozambique'.

28 Maria Luisa Norton Pinto Teixeira, 'The railways of Mozambique: A regional or colonial project? 1895–1950', master's thesis, Concordia University, Montreal, 1991.

29 Alden, *Mozambique and the Construction*.

30 Pinto Teixeira, 'The railways of Mozambique'.

31 Pereira, 'Colonial railways of Mozambique'.

32 Leroy Vail, 'The making of an imperial slum: Nyasaland and its railways, 1895–1935', *Journal of African History*, vol. 16, no. 1 (1975), pp. 89–112.

33 Pereira, 'Colonial railways of Mozambique'.

34 Diogo and Navarro, 'Re-designing Africa'.

35 Alden, *Mozambique and the Construction*.

36 Smith, 'The idea of Mozambique'.

37 Diogo and Navarro, 'Re-designing Africa'; Robert Nkana, 'Malawi

railways: An historical review', *Society of Malawi Journal*, vol. 52, no. 1 (1999), pp. 39–45.

38 Sayaka Funada-Classen, *The Origins of War in Mozambique: A History of Unity and Division*, trans. Masako Osada (Somerset West, African Minds, 2013).

39 Kirshner and Baptista, 'Corridors as empty signifiers'.

40 Tina Sideris, 'War, gender and culture: Mozambican women refugees', *Social Science & Medicine*, vol. 56, no. 4 (2003), pp. 713–24.

41 Tilman Brück, 'War and reconstruction in northern Mozambique', *Economics of Peace and Security Journal*, vol. 1, no. 1 (2006), pp. 30–9.

42 David Alexander Robinson, 'Curse on the land: A history of the Mozambican Civil War', PhD thesis, University of Western Australia, Perth, 2006.

43 Horace Campbell, 'War, reconstruction and dependence in Mozambique', *Third World Quarterly*, vol. 6, no. 4 (1984), pp. 839–67.

44 Scholvin and Plagemann, 'Transport infrastructure'.

45 James Derrick Sidaway, 'Mozambique: Destabilization, state, society and space', *Political Geography*, vol. 11, no. 3 (1992), pp. 239–58.

46 James Derrick Sidaway and Marcus Power, 'Sex and violence on the wild frontiers: The aftermath of state socialism in the periphery', in John Pickles and Adrian Smith (eds), *Theorising Transition: The Political Economy of Post-Communist Transformations* (London, Routledge, 1998).

47 Alden, *Mozambique and the Construction*.

48 William Minter, *Apartheid's Contras: An Inquiry into the Roots of War in Angola and Mozambique* (London, Zed Books, 1994).

49 Marina Ottaway, 'Mozambique: From symbolic socialism to symbolic reform', *Journal of Modern African Studies*, vol. 26, no. 2 (1988), pp. 211–26.

50 Alden, *Mozambique and the Construction*.

51 African Elections Database, 'Elections in Mozambique', https://africanelections.tripod.com/mz.html.

52 Inter-Parliamentary Union, 'Mozambique Parliamentary Chamber: Assembleia da Republica: Elections Held in 1994', http://archive.ipu.org/parline-e/reports/arc/2223_94.htm.

53 Sandra Sequiera, 'Transport costs and firm behaviour: Evidence from Mozambique and South Africa', in Olivier Cadot et al. (eds), *Where to Spend the Next Million? Applying Impact Evaluation to Trade*

Assistance (Washington, DC, International Bank for Reconstruction and Development/World Bank, 2011), pp. 123–61.

54 Logistics Cluster, 'Mozambique railway assessment', 2022, https://dlca. logcluster.org/24-mozambique-railway-assessment.

55 Larry Phipps, *Technical Assessment Report: The Concessioning of Mozambique Railways: Rapid Support Option* (Regional Activity to Promote Integration through Dialogue and Policy Implementation (RAPID), Gaborone, 2001), https://pdf.usaid.gov/pdf_docs/Pnact575.pdf.

56 Reuters, 'US Railnet plans $11 billion railway project in southern Africa', 9 March 2020, www.reuters.com/article/idUSKBN20V0OV/; AllAfrica, 'Mozambique to introduce express train to South Africa and Eswatini', 5 September 2022, https://allafrica.com/stories/202209050390.html.

57 Carolina Dominguez-Torres and Cecilia Briceño-Garmendia, *Mozambique's Infrastructure: A Continental Perspective* (Washington, DC, International Bank for Reconstruction and Development/World Bank, 2011).

58 Scholvin and Plagemann, 'Transport infrastructure'.

59 Agnieszka Flak and Marina Lopes, 'Analysis: Poor railways, ports put brake on Mozambique's coal rush', Reuters, 16 April 2013.

60 Geoffrey Wood and Pauline Dibben, 'Ports and shipping in Mozambique: Current concerns and policy options', *Maritime Policy & Management*, vol. 32, no. 2 (2005), pp. 139–57; Glen Robbins and David Perkins, 'Mining FDI and infrastructure development on Africa's east coast: Examining the recent experience of Tanzania and Mozambique', *Journal of International Development*, vol. 24 (2012), pp. 220–36.

61 Flak and Lopes, 'Analysis: Poor railways'.

62 Silvia Antonioli and Jim Regan, 'Rio Tinto pulls plug on ill-fated Mozambique coal venture', Reuters, 30 July 2014.

63 Thomas Selemane, 'From Vale to Vulcan: "New wine in old bottles?" An analysis of Vale's exit from Mozambique', *Extractive Industries and Society*, vol. 15 (2023), 101302.

64 Ayokunu Adedokun, 'Emerging challenges to long-term peace and security in Mozambique', *Journal of Social Encounters*, vol. 1, no. 1 (2017), pp. 37–53.

65 Kirshner and Baptista, 'Corridors as empty signifiers'.

3. Convenience

1 Theodore Roosevelt, *Message of the President of the United States*

Communicated to the Two Houses of Congress at the Beginning of the First Session of the Fifty-Seventh Congress (Washington, DC, Government Printing Office, 1901), p. 27.

2   Justin Harper, 'Suez blockage is holding up $9.6bn of goods a day', BBC News, 26 March 2021, www.bbc.com/news/business-56533250.

3   United Nations Conference on Trade and Development (UNCTAD), *Review of Maritime Transport 2023: Towards a Green and Just Transition* (New York, United Nations, 2023), https://unctad.org/system/files/official-document/rmt2023_en.pdf.

4   Matthew Parker, *Panama Fever: The Epic Story of the Building of the Panama Canal* (New York, Anchor, 2009).

5   Noel Maurer and Carlos Yu, *The Big Ditch: How America Took, Built, Ran, and Ultimately Gave Away the Panama Canal* (Princeton, NJ, Princeton University Press, 2011).

6   Willis Fletcher Johnson, *Four Centuries of the Panama Canal* (New York, Henry Holt, 1906).

7   David McCullough, *The Path between the Seas: The Creation of the Panama Canal, 1870–1914* (New York, Simon & Schuster, 1977).

8   José Carlos Rodrigues, *The Panama Canal: Its History, Its Political Aspects, and Financial Difficulties* (London, Sampson Low, 1885).

9   McCullough, *The Path*.

10 Brantz von Mayer, 'Dredging the Panama Canal', *Dredge Brokers*, https://dredgebrokers.com/dredging-the-panama-canal/.

11 Johnson, *Four Centuries*.

12 J. Saxon Mills, *The Panama Canal: A History and Description of the Enterprise* (New York, Sully & Kleinteich, 1913).

13 Maurer and Yu, *The Big Ditch*.

14 Parker, *Panama Fever*.

15 Johnson, *Four Centuries*.

16 'Gustave Eiffel', www.toureiffel.paris/en/the-monument/gustave-eiffel.

17 Johnson, *Four Centuries*.

18 Maurer and Yu, *The Big Ditch*.

19 Parker, *Panama Fever*.

20 Logan Marshall, *The Story of the Panama Canal* (Philadelphia, PA, John C. Winston, 1913).

21 Lawrence A. Clayton, 'The Nicaragua canal in the nineteenth century: Prelude to American empire in the Caribbean', *Journal of Latin American Studies*, vol. 19, no. 2 (1987), pp. 323–52.

22 Jackson Crowell, 'The United States and a Central American canal,

1869–1877', *Hispanic American Historical Review*, vol. 49, no. 1 (1969), pp. 27–52.

23 Alexander Missal, *Seaway to the Future: American Social Visions and the Construction of the Panama Canal* (Madison, University of Wisconsin Press, 2008).

24 Mills, *The Panama Canal*.

25 Fernando Manfredo Jr, 'The future of the Panama Canal', *Journal of Interamerican Studies and World Affairs*, vol. 35, no. 3 (1993), pp. 103–28; C. Willard Hayes, 'The Nicaragua Canal Route', *Science*, vol. 10, no. 239 (1899), pp. 97–104.

26 Clayton, 'The Nicaragua canal'.

27 K. Rockwell, 'A brief history of the Panama Canal', *Professional Memoirs, Corps of Engineers, United States Army, and Engineer Department at Large*, vol. 1, no. 2 (1909), pp. 164–74.

28 Francisco J. Montero Llácer, 'Panama Canal management', *Marine Policy*, vol. 29 (2005), pp. 25–37.

29 Clayton, 'The Nicaragua canal'.

30 Charles D. Ameringer, 'The Panama Canal lobby of Philippe Bunau-Varilla and William Nelson Cromwell', *American Historical Review*, vol. 68, no. 2 (1963), pp. 346–63.

31 Clayton, 'The Nicaragua canal'.

32 Lesley A. DuTemple, *The Panama Canal* (Minneapolis, MN, Lerner, 2003).

33 Stephen G. Rabe, 'Theodore Roosevelt, the Panama Canal, and the Roosevelt Corollary to the Monroe Doctrine', *Idées d'Amériques (ideAs)*, vol. 22 (2023).

34 Missal, *Seaway to the Future*.

35 Theodore Roosevelt, quoted in Stephen J. Randall, *Colombia and the United States: Hegemony and Independence* (Athens, University of Georgia Press, 1992, p. 85).

36 Johnson, *Four Centuries*.

37 Ameringer, 'The Panama Canal lobby'; Charles D. Ameringer, 'Philippe Bunau-Varilla: New light on the Panama Canal Treaty', *Hispanic American Historical Review*, vol. 46, no. 1 (1966), pp. 28–52.

38 Panama Canal Treaty, 1903, www.historycentral.com/HistoricalDocuments/PanamaCanalTreaty.html.

39 Roosevelt, in Randall, *Colombia and the United States*, pp. 85–6.

40 Julie Greene, *The Canal Builders: Making America's Empire at the Panama Canal* (New York, Penguin Press, 2009).

41 Missal, *Seaway to the Future*.

42 Robert N. Wiedenmann and J. Ray Fisher, *The Silken Thread: Five Insects and Their Impacts on Human History* (Oxford, Oxford University Press, 2021).

43 Enrique Chaves-Carballo, 'Ancon Hospital: An American hospital during the construction of the Panama Canal, 1904–1914', *Military Medicine*, vol. 164, no. 10 (1999), pp. 725–30.

44 Wiedenmann and Fisher, *The Silken Thread*; Maurer and Yu, *The Big Ditch*.

45 Greene, *The Canal Builders*.

46 Rockwell, 'A brief history'.

47 Greene, *The Canal Builders*.

48 Ashley Carse, *Beyond the Big Ditch: Politics, Ecology, and Infrastructure at the Panama Canal* (Cambridge, MA, MIT Press, 2014).

49 Maurer and Yu, *The Big Ditch*; https://ufdcimages.uflib.ufl.edu/AA/00/07/66/20/00001/PanamaCanalMuseum.pdf.

50 Greene, *The Canal Builders*.

51 von Mayer, 'Dredging'.

52 Missal, *Seaway to the Future*.

53 Julie Greene, 'Spaniards on the silver roll: Labor troubles and liminality in the Panama Canal Zone, 1904–1914', *International Labor and Working-Class History*, vol. 66 (2004), pp. 78–98.

54 'Oldtimers well remember the Yellow Peril', *Panama Canal Review*, 2 April 1954, pp. 8–9, www.govinfo.gov/content/pkg/GOVPUB-W79-f1858e8bdeb9f31c7a08f2c8fce156fe/pdf/GOVPUB-W79-f1858e8bdeb9f31c7a08f2c8fce156fe.pdf.

55 Frederic J. Haskin, *The Panama Canal* (Garden City, NY, Doubleday, Page & Company, 1914).

56 *Official Handbook of the Panama Canal* (Ancon, Secretary of the Isthmian Canal Commission, 1913), www.govinfo.gov/content/pkg/GOVPUB-Y3_IS7-7beac19afbe1a44295af86691290aa27/pdf/GOVPUB-Y3_IS7-7beac19afbe1a44295af86691290aa27.pdf.

57 Autoridad del Canal de Panamá, 'Culebra Cut', 2024, https://pancanal.com/en/culebra-cut.

58 Greene, 'Spaniards on the silver roll'.

59 Caroline Lieffers, 'The Panama Canal's forgotten casualties', *The Conversation*, 16 April 2018, https://theconversation.com/the-panama-canals-forgotten-casualties-93536.

60 Noel Maurer and Carlos Yu, 'What T. R. took: The economic impact of

the Panama Canal, 1903–1937', *Journal of Economic History*, vol. 68, no. 3 (2008), pp. 686–721.

61 Robert M. Brown, 'Five years of the Panama Canal: An evaluation', *Geographical Review*, vol. 9, no. 3 (1920), pp. 191–8; Maurer and Yu, *The Big Ditch*.

62 G. G. Huebner, 'Economic aspects of the Panama Canal', *American Economic Review*, vol. 5, no. 4 (1915), pp. 816–29.

63 Martin B. Travis and James T. Watkins, 'Control of the Panama Canal: An obsolete shibboleth?' *Foreign Affairs*, vol. 37, no. 3 (1959), pp. 407–18.

64 Greene, *The Canal Builders*.

65 Maurer and Yu, *The Big Ditch*.

66 Manfredo, 'The future of the Panama Canal'.

67 Maurer and Yu, 'What T. R. took'.

68 Marixa Lasso, 'Citizens to "natives": Tropical politics of depopulation at the Panama Canal Zone', *Environmental History*, vol. 21, no. 2 (2016), pp. 240–9.

69 Ashley Carse, 'Nature as infrastructure: Making and managing the Panama Canal watershed', *Social Studies of Science*, vol. 42, no. 4 (2012), pp. 539–63.

70 Samuel Cox, 'H-057-2: *I-400* and Operation Cherry Blossoms at night: Japanese plan for biological warfare – September 1945', Naval History and Heritage Command, www.history.navy.mil/about-us/leadership/director/directors-corner/h-grams/h-gram-057/h-057-2.html.

71 Carse, *Beyond the Big Ditch*.

72 Maurer and Yu, *The Big Ditch*.

73 Alan McPherson, 'Courts of world opinion: Trying the Panama flag riots of 1964', *Diplomatic History*, vol. 28, no. 1 (2004), pp. 83–112.

74 Michael Donoghue, *Borderland on the Isthmus: Race, Culture, and the Struggle for the Canal Zone* (Durham, NC, Duke University Press, 2014).

75 Alan McPherson, 'From "punks" to geopoliticians: US and Panamanian teenagers and the 1964 Canal Zone riots', *Americas*, vol. 58, no. 3 (2002), pp. 395–418.

76 *Report on the Events in Panama, January 9–12, 1964* (Geneva, International Commission of Jurists, 1964).

77 McPherson, 'Courts of world opinion'.

78 Donoghue, *Borderland on the Isthmus*.

79 Rebecca L. Grant, *Operation Just Cause and the US Policy Process* (Santa Monica, CA, RAND, 1991).

80 Katarzyna Kwiatkowska, 'Manuel Noriega and his impact on the events in Panama from 1981 to 1989', *Ameryka Łacińska*, vol. 30, no. 2 (2022), pp. 25–48.

81 Grant, *Operation Just Cause.*

82 Maurer and Yu, *The Big Ditch.*

83 'Fighting in Panama: The President; A transcript of Bush's address on the decision to use force in Panama', *New York Times*, 21 December 1989.

84 Ronald H. Cole, *Operation Just Cause: The Planning and Execution of Joint Operations in Panama, February 1988 – January 1990* (Washington, DC, Joint History Office, Office of the Chairman of the Joint Chiefs of Staff, 1995); United States Southern Command (US SOUTHCOM), *Public Affairs after Action Report Supplement, 'Operation Just Cause', Dec. 20, 1989–Jan. 31, 1990*, https://nsarchive2.gwu.edu/nsa/ DOCUMENT/950206.htm; Carl T. Bogus, 'The invasion of Panama and the rule of law', *International Lawyer*, vol. 26, no. 3 (1992), pp. 781–7.

85 Ricardo Hausmann, Miguel Angel Santos and Juan Obach, *Appraising the Economic Potential of Panama: Policy Recommendations for Sustainable and Inclusive Growth*, CID faculty working paper 334, 2017, https://growthlab.hks.harvard.edu/sites/projects.iq.harvard.edu/files/ growthlab/files/panama_policy_wp_334.pdf.

86 Manfredo, 'The future of the Panama Canal'.

87 Autoridad del Canal de Panamá, 'One million and counting: Panama Canal Authority commemorates history-making transit through canal', 14 October 2010, https://pancanal.com/en/ one-million-and-counting-panama-canal-authority-commemorates- history-making-transit-through-canal.

88 Adriana Tapia Zafra and Sarah Holder, 'The world faces a $270 billion traffic jam in Panama', Bloomberg, 12 January 2024, www.bloomberg. com/news/articles/2024-01-12/how-the-panama-canal-s-drought-effects- global-trade-big-take-podcast.

89 Mariner Wang, 'The role of Panama Canal in global shipping', *Maritime Business Review*, vol. 2, no. 3 (2017), pp. 247–60.

90 Joyendu Bhadury, 'Panama Canal expansion and its impact on East and Gulf coast ports of USA', *Maritime Policy & Management*, vol. 43, no. 8 (2016), pp. 928–44.

91 Francisco J. Montero Llácer, 'The Panama Canal: Operations and traffic', *Marine Policy*, vol. 29 (2005), pp. 223–34.

92 Walt Bogdanovich, Jacqueline Williams and Ana Graciela Méndez, 'The New Panama Canal: A risky bet', *New York Times*, 22 June 2016.

93 Wang, 'The role'.

94 Rodolfo Sabonge, *The Panama Canal Expansion: A Driver of Change for Global Trade Flows* (Santiago, United Nations Economic Commission for Latin America and the Caribbean, 2014).

95 Thi Yen Pham, Ki Young Kim and Gi-Tae Yeo, 'The Panama Canal expansion and its impact on east–west liner shipping route selection', *Sustainability*, vol. 10 (2018), 4353.

96 Autoridad del Canal de Panamá, 'Advisory to shipping no. A-27-2024', 9 August 2024; Wang, 'The role'; Marsea Nelson, 'Ten facts about the Panama Canal', World Wildlife Fund (WWF), 27 September 2010, www.worldwildlife.org/blogs/good-nature-travel/posts/ten-facts-about-the-panama-canal.

97 Autoridad del Canal de Panamá, 'Top 15 countries by origin and destination of cargo FY 2023', 2024, https://pancanal.com/en/statistics.

98 Sabonge, *The Panama Canal Expansion*; Pham, Kim and Yeo, 'The Panama Canal expansion'.

99 Carse, *Beyond the Big Ditch*.

100 Greta Rosen Fondahn, 'Maersk to use rail for some vessels to bypass Panama Canal amid drought', Reuters, 11 January 2024, www.reuters.com/business/maersk-bypass-panama-canal-amid-drought-2024-01-11.

101 Ruth Liao and Bloomberg, 'Panama Canal has gotten so dry and backed up after brutal drought that shippers are paying up to $4m to jump the queue', *Fortune*, 4 December 2023, https://fortune.com/2023/12/04/panama-canal-dry-backed-up-brutal-drought-shippers-paying-4m-jump-queue.

102 Peter Eavis, 'Drought saps the Panama Canal, disrupting global trade', *New York Times*, 1 November 2023.

103 Mia Jankowicz, 'The Panama Canal is so clogged up that a shipping company paid $4 million to jump the line: Report', *Business Insider*, 9 November 2023, www.businessinsider.com/panama-canal-shipping-company-paid-record-4-million-skip-line-2023-11.

104 NASA Earth Observatory, 'Heavy rains in Central and South America', 12 December 2010, https://earthobservatory.nasa.gov/images/47745/heavy-rains-in-central-and-south-america.

105 Wang, 'The role'.

106 'Nicaragua cancels Chinese plan for controversial canal 10 years on', *Guardian*, 8 May 2024, www.theguardian.com/world/article/2024/

may/08/nicaragua-cancel-china-canal; Simon Romero, 'China's ambitious rail projects crash into harsh realities in Latin America', *New York Times*, 3 October 2015.

107 'Panama Canal expects new water reservoir for ship crossings in 6 years', Reuters, 8 July 2024, www.reuters.com/markets/commodities/panama-canal-expects-new-water-reservoir-ship-crossings-6-years-2024-07-08.

108 Guido Bilbao, 'Panamanian indigenous people act to protect the forest from invading loggers', *Mongabay*, 2 April 2019, https://news.mongabay.com/2019/04/panamanian-indigenous-people-act-to-protect-the-forest-from-invading-loggers.

109 Carse, 'Nature as infrastructure'.

110 Stanley Heckadon Moreno, 'Impact of development on the Panama Canal environment', *Journal of Interamerican Studies and World Affairs*, vol. 35, no. 3 (1993), pp. 129–49; Carse, 'Nature as infrastructure'.

111 Marixa Lasso, *Erased: The Untold Story of the Panama Canal* (Cambridge, MA, Harvard University Press, 2019).

112 Carse, *Beyond the Big Ditch*.

113 Christopher Cairns, 'China's investment setbacks in Panama', *Diplomat*, 26 February 2022, https://thediplomat.com/2022/02/chinas-investment-setbacks-in-panama.

114 Mat Youkee, 'Panama the new flashpoint in China's growing presence in Latin America', *Guardian*, 28 November 2018, www.theguardian.com/world/2018/nov/28/panama-china-us-latin-america-canal.

115 Daniel F. Runde and Amy Doring, 'Key decision point coming for the Panama Canal', *Center for Strategic & International Studies (CSIS)*, 21 May 2021, www.csis.org/analysis/key-decision-point-coming-panama-canal.

116 Parker, *Panama Fever*.

117 Greene, *The Canal Builders*.

4. Reimagination

1 'HRH Crown Prince Mohammed bin Salman announces designs for THE LINE, the city of the future in NEOM', 25 July 2022, www.neom.com/en-us/newsroom/hrh-announces-theline-designs.

2 Alistair Walsh, 'Saudi Arabia grants robot citizenship', *DW*, 28 October 2017, www.dw.com/en/saudi-arabia-grants-citizenship-to-robot-sophia/a-41150856; Lama Alhamawi, 'How a Saudi healthcare startup

is using AI to transform the diagnosis of chronic diseases', *Arab News*, 15 June 2024, www.arabnews.com/node/2511661/saudi-arabia.

3 'WTTC commends Saudi Arabia on launch of tourism investment enablers program', World Travel & Tourism Council (WTTC), 5 March 2024, https://wttc.org/news-article/wttc-commends-saudi-arabia-on-launch-of-tourism-investment-enablers-program.

4 Alshimaa Aboelmakarem Farag, 'The story of NEOM city: Opportunities and challenges', in Sahar Attia, Zeinab Shafik and Asmaa Ibrahim (eds), *New Cities and Community Extensions in Egypt and the Middle East: Visions and Challenges* (Cham, Springer, 2019), pp. 35–49.

5 'THE LINE: The future of urban living', www.neom.com/en-us/regions/theline.

6 'THE LINE: Saudi Arabia's city of the future in NEOM', 2023, www.youtube.com/watch?v=0amD9QoTH9M&ab_channel=DiscoveryUK.

7 Nadia Yusuf and Dareen Abdulmohsen, 'Saudi Arabia's NEOM project as a testing ground for economically feasible planned cities: Case study', *Sustainability*, vol. 15, no. 1 (2023), p. 608.

8 NEOM, 'THE LINE'.

9 Ibid.

10 Giles Pendleton, 'THE LINE: Taking vertical urbanism to the next level', www.neom.com/en-us/regions/theline/vertical-urbanism.

11 'What is NEOM?', www.neom.com/en-us/about.

12 'Neom: Tourism experiences of the future', UN Tourism, www.unwto.org/neom-tourism-experiences-of-the-future; NEOM, 'THE LINE'.

13 Zainab Fattah and Matthew Martin, 'Saudis scale back ambition for $1.5 trillion desert project Neom', Bloomberg, 5 April 2024, www.bloomberg.com/news/articles/2024-04-05/saudis-scale-back-ambition-for-1-5-trillion-desert-project-neom.

14 'The Line: City of the future in Saudi Arabia', https://line-neom.com.

15 'NEOM Green Hydrogen Company completes financial close at a total investment value of USD 8.4 billion in the world's largest carbon-free green hydrogen plant', 22 May 2023, www.neom.com/en-us/newsroom/neom-green-hydrogen-investment.

16 NEOM, 'THE LINE'.

17 Merlyn Thomas and Vibeke Venema, 'Neom: What's the green truth behind a planned eco-city in the Saudi desert?', BBC News, 21 February 2022, www.bbc.com/news/blogs-trending-59601335.

18 'Saudi Arabia's NEOM city enlists Dutch greenhouse company to grow crops in desert', *Al Arabiya*, 8 August 2023, https://english.alarabiya.

net/News/saudi-arabia/2023/08/08/Saudi-Arabia-s-NEOM-city-enlists-Dutch-greenhouse-company-to-grow-crops-in-desert.

19 Michael Batty, 'The Linear City: Illustrating the logic of spatial equilibrium', *Computational Urban Science*, vol. 2, no. 8 (2022).

20 NEOM, 'THE LINE'.

21 Pendleton, 'THE LINE'.

22 Ibid.

23 Alvin R. Cabral, 'Saudi Arabia's Neom teams up with Germany's Volocopter for urban air mobility system', *National*, 1 December 2021, www.thenationalnews.com/business/aviation/2021/12/01/saudi-arabias-neom-teams-up-with-germanys-volocopter-for-urban-air-mobility-system.

24 Justin Scheck, Rory Jones and Summer Said, 'A prince's $500 billion desert dream: Flying cars, robot dinosaurs and a giant artificial moon', *Wall Street Journal*, 25 July 2019.

25 Simon Sadler, *Archigram: Architecture without Architecture* (Cambridge, MA, MIT Press, 2005).

26 Kent E. Portney, *Taking Sustainable Cities Seriously: Economic Development, the Environment, and Quality of Life in American Cities*, second edition (Cambridge, MA, MIT Press, 2013).

27 Batty, 'The Linear City'.

28 Natalia E. Paszkowska-Kaczmarek, '*The Line*: The Saudi-Arabian linear city concept as the prototype of future cities', *Architecturae et Artibus*, vol. 13, no. 2 (2021), pp. 33–46.

29 George R. Collins, 'Linear planning throughout the world', *Journal of the Society of Architectural Historians*, vol. 18, no. 3 (1959), pp. 74–93.

30 Paszkowska-Kaczmarek, '*The Line*'; Batty, 'The Linear City'.

31 Jennifer Gray, 'Reading Broadacre', Frank Lloyd Wright Foundation, 1 October 2018, https://franklloydwright.org/reading-broadacre.

32 Gus Lubin, 'Why architect Le Corbusier wanted to demolish downtown Paris', *Business Insider*, 20 August 2013, www.businessinsider.com/le-corbusiers-plan-voisin-for-paris-2013-7.

33 Batty, 'The Linear City'; Kurt Kohlstedt, 'Ville Radieuse: Le Corbusier's functionalist plan for a utopian "Radiant City"', *99% Invisible*, 23 February 2018, https://99percentinvisible.org/article/ville-radieuse-le-corbusiers-functionalist-plan-utopian-radiant-city.

34 Peter Hall, *Cities of Tomorrow: An Intellectual History of Urban Planning and Design since 1880*, fourth edition (Chichester, Wiley-Blackwell, 2014).

35   Brian Ackley, 'Le Corbusier's Algerian fantasy: Blocking the Casbah',
      Bidoun (2006), www.bidoun.org/articles/le-corbusier-s-algerian-fantasy;
      Collins, 'Linear planning'.

36   Collins, 'Linear planning'.

37   Alan Smart, 'Beyond utopia: Representing life in the productivist city',
      Architecture and Culture, vol. 3, no. 3 (2015), pp. 297–313.

38   Anna Zaręba et al., 'Linear cities as an alternative for the sustainable
      transition of urban areas in harmony with natural environment
      principles', in Francesco Alberti et al. (eds), *Urban and Transit Planning:
      Towards Liveable Communities: Urban Places and Design Spaces*, second
      edition (Cham, Springer, 2022), pp. 87–99.

39   Collins, 'Linear planning'; Zaręba et al., 'Linear cities'.

40   Eir Nolsøe, 'Why Mohammed bin Salman has been forced to rein in his
      dreams of a mirror city', *Daily Telegraph*, 9 April 2024.

41   'Urban development', World Bank, 2023, www.worldbank.org/en/topic/
      urbandevelopment/overview.

42   *Kingdom of Saudi Arabia: Report on the Implementation of the
      New Urban Agenda* (Riyadh, Ministry of Municipal, Rural Affairs &
      Housing, 2022), www.urbanagendaplatform.org/sites/default/files/2023-
      04/NUA%20Report%20Final_05Dec2022-compressed.pdf.

43   Nat Barker, 'Sustainability and liveability claims of Saudi 170-kilometre
      city are "naïve" say experts', *Dezeen*, 8 August 2022, www.dezeen.
      com/2022/08/08/sustainability-liveability-the-line-saudi-170km-city-naive.

44   Alain Musset, 'Neom et *The Line* (Arabie saoudite) utopie futuriste
      ou cauchemar urbain?', *L'Information géographique*, vol. 87 (2023),
      pp. 139–61.

45   Steven Griffiths and Benjamin K. Sovacool, 'Rethinking the future low-
      carbon city: Carbon neutrality, green design, and sustainability tensions
      in the making of Masdar City', *Energy Research & Social Science*, vol. 62
      (2020), 101368; Jamey Keaten, 'On sidelines of COP28, Emirati "green
      city" falls short of ambitions, but still delivers lessons', Associated Press,
      8 December 2023, https://apnews.com/article/masdar-city-abu-dhabi-
      cop28-sustainable-city-8d656e09d1cdc91ea08842ca332809b2.

46   Musset, 'Neom et *The Line*'.

47   Barker, 'Sustainability and liveability'.

48   NEOM, 'THE LINE'.

49   Rafael Prieto-Curiel and Dániel Kondor, 'Arguments for building The
      Circle and not The Line in Saudi Arabia', *npj Urban Sustainability*,
      vol. 3, no. 35 (2023).

50 Laura Tizzo, 'Avião ou borboleta? Entenda as inspirações de Lúcio Costa para o projeto de Brasília', *Globo*, 4 June 2019, https://g1.globo.com/df/distrito-federal/noticia/2019/06/04/aviao-ou-borboleta-entenda-as-inspiracoes-de-lucio-costa-para-o-projeto-de-brasilia.ghtml.

51 Prieto-Curiel and Kondor, 'Arguments for building The Circle'.

52 Batty, 'The Linear City'.

53 Oz Hassan, 'Artificial intelligence, Neom and Saudi Arabia's economic diversification from oil and gas', *Political Quarterly*, vol. 91, no. 1 (2020), pp. 222–7; Gregg Carlstrom, 'The power vacuum in the Middle East', *Foreign Affairs*, 6 March 2024.

54 Ruth Michaelson, '"It's being built on our blood": The true cost of Saudi Arabia's $500bn megacity', *Guardian*, 4 May 2020.

55 'Saudi Arabia: ILO forced labor complaint a wake-up call', Human Rights Watch, 5 June 2024, www.hrw.org/news/2024/06/05/saudi-arabia-ilo-forced-labor-complaint-wake-call.

56 'Saudi Arabia: Labor reforms insufficient', Human Rights Watch, 25 March 2021, www.hrw.org/news/2021/03/25/saudi-arabia-labor-reforms-insufficient.

57 'Saudi security forces flatten old quarter of Awamiya', Al Jazeera, 9 August 2017, www.aljazeera.com/news/2017/8/9/saudi-security-forces-flatten-old-quarter-of-awamiya.

58 'The Jeddah demolitions: Forced evictions and neighborhood destruction put more than a million residents at risk', *DAWN*, 17 October 2022, https://dawnmena.org/the-jeddah-demolitions-forced-evictions-and-neighborhood-destructions-put-more-than-a-million-residents-at-risk.

59 Quoted in *Discovery UK*, 'THE LINE'.

60 *The Dark Side of Neom: Expropriation, Expulsion and Prosecution of the Region's Inhabitants*, ALQST for Human Rights, 2023, https://alqst.org/uploads/the-dark-side-of-neom-expropriation-expulsion-and-prosecution-en.pdf.

61 'Countries and territories', Freedom House, 2024, https://freedomhouse.org/countries/freedom-world/scores; 'GDP per capita, PPP (current international $)', World Bank, 2022, https://data.worldbank.org/indicator/NY.GDP.PCAP.PP.CD?most_recent_value_desc=true&year_high_desc=true.

62 Saudatu Bah, 'Dezeen Agenda features "moral dilemma" facing architects designing Neom in Saudi Arabia', *Dezeen*, 20 December 2022, www.dezeen.com/2022/12/20/peter-frankental-interview-saudi-arabia-neom-project-dezeen-agenda.

63 Max Jeffrey, 'The strange tale of NEOM: Saudi Arabia's struggling

desert megacity', *Spectator*, 24 April 2021, www.spectator.co.uk/article/saudi-s-city-on-hold.

64 Nolsøe, 'Why Mohammed bin Salman'.

65 Michio Kaku, in *The Placemakers: Episode 2 – THE LINE*, on NEOM, 'The Line' webpage.

66 Musset, 'Neom et *The Line*'.

5. Resistance

1 'The Baltic Way', *Congressional Record: Senate*, 19 September 1989, p. 20781, www.govinfo.gov/content/pkg/GPO-CRECB-1989-pt15/pdf/GPO-CRECB-1989-pt15-5-2.pdf.

2 S. C. Rowell, *Lithuania Ascending: A Pagan Empire within East-Central Europe, 1295–1345* (Cambridge, Cambridge University Press, 1994).

3 Janis Paliepa, *The Origin of the Baltic and Vedic Languages: Baltic Mythology* (Bloomington, IN, AuthorHouse, 2011); Scott Spires, 'Lithuanian linguistic nationalism and the cult of antiquity', *Nations and Nationalism*, vol. 5, no. 4 (1999), pp. 485–500.

4 Reinhard Krumm, Tõnis Stamberg and Irina Strapatšuk, *Feeling Cornered: An Analysis of the Russian-Speaking Minority in Estonia* (Tallinn, Friedrich-Ebert-Stiftung, 2023); Anchal Vohra, 'Latvia is going on offense against Russian culture', *Foreign Policy*, 21 March 2023, https://foreignpolicy.com/2023/03/21/latvia-is-going-on-offense-against-russian-culture.

5 Toomas Hendrik Ilves, 'Välisminister Ilvese loeng Rootsi Välispoliitika Instituudis: "Eesti kui Põhjamaa" (inglise keeles)', *Välisministeerium*, 14 December 1999, https://web.archive.org/web/20010260801806/https://vm.ee/et/uudised/valisminister-ilvese-loeng-rootsi-valispoliitika-instituudis-eesti-kui-pohjamaa-inglise; Kari Alenius, '"Let us be like the Finns!": The image of Finland and the Finns as a key element in the construction of Estonian national identity', in Steven G. Ellis and Lud'a Klusáková (eds), *Imagining Frontiers, Contesting Identities* (Pisa, Edizioni Plus–Pisa University Press, 2007), pp. 377–91.

6 Imtiaz Khan and Ali Shahaab, 'Estonia is a "digital republic": What that means and why it may be everyone's future', *Conversation*, 7 October 2020, https://theconversation.com/estonia-is-a-digital-republic-what-that-means-and-why-it-may-be-everyones-future-145485; Education Estonia, 'How it all began? From Tiger Leap to digital society', www.educationestonia.org/tiger-leap; 'Estonian Internet voting',

European Commission, 29 July 2019, https://ec.europa.eu/digital-building-blocks/sites/pages/viewpage.action?pageId=533365949.

7 'Baltic song and dance celebrations', UNESCO, https://ich.unesco.org/en/RL/baltic-song-and-dance-celebrations-00087.

8 'Sutartinės, Lithuanian multipart songs', UNESCO, https://ich.unesco.org/en/RL/sutartins-lithuanian-multipart-songs-00433.

9 Aldis Purs, *Baltic Facades: Estonia, Latvia and Lithuania since 1945* (London, Reaktion, 2012).

10 Geoffrey Roberts, *The Soviet Union and the Origins of the Second World War: Russo-German Relations and the Road to War, 1933–1941* (New York, St Martin's Press, 1995).

11 William R. Trotter, *A Frozen Hell: The Russo-Finnish Winter War of 1939–1940* (Chapel Hill, NC, Algonquin, 1991).

12 Jennifer A. Yoder, *World War II Memory and Contested Commemorations in Europe and Russia* (Oxford, Oxford University Press, 2024).

13 Aleksandras Shtromas, 'The Baltic states as Soviet republics: Tensions and contradictions', in Graham Smith (ed.), *The Baltic States: The National Self-Determination of Estonia, Latvia and Lithuania* (New York, St Martin's Press, 1994), pp. 86–117; Lauri Mälksoo, *Illegal Annexation and State Continuity: The Case of the Incorporation of the Baltic States by the USSR*, second edition (Leiden, Brill Nijhoff, 2003).

14 Rain Liivoja, 'Competing histories: Soviet war crimes in the Baltic states', in Kevin Jon Heller and Gerry Simpson (eds), *The Hidden Histories of War Crimes Trials* (Oxford, Oxford University Press, 2013), pp. 248–66.

15 John Hiden and Patrick Salmon, *The Baltic Nations and Europe: Estonia, Latvia and Lithuania in the Twentieth Century*, revised edition (London, Longman, 1994).

16 Aigi Rahi-Tamm and Irena Salēniece, 'Re-educating teachers: Ways and consequences of Sovietization in Estonia and Latvia (1940–1960) from the biographical perspective', *Journal of Baltic Studies*, vol. 47, no. 4 (2016), pp. 451–72.

17 David Puderbaugh, 'How choral music saved a nation: The 1947 Estonian National Song Festival and the song festivals of Estonian's [*sic*] Soviet occupation', *Choral Journal*, vol. 49, no. 4 (2008), pp. 28–43.

18 Guntis Šmidchens, *The Power of Song: Nonviolent National Culture in the Baltic Singing Revolution* (Seattle, University of Washington Press, 2014); Isabelle T. Kreindler, 'Baltic area languages in the Soviet Union:

A sociolinguistic perspective', *Journal of Baltic Studies*, vol. 19, no. 1 (1988), pp. 5–20.

19 Michael Loader, 'The rebellious republic: The 1958 education reform and Soviet Latvia', *Latvijas Vēstures Institūta Žurnāls*, vol. 100, no. 3 (2016), pp. 113–39; Michael Loader, 'Restricting Russians: Language and immigration laws in Soviet Latvia, 1956–1959', *Nationalities Papers*, vol. 45, no. 6 (2017), pp. 1082–99.

20 Michael Loader, 'A Stalinist purge in the Khrushchev era? The Latvian Communist Party purge, 1959–1963', *Slavonic and East European Review*, vol. 96, no. 2 (2018), pp. 244–82; Daniel Baldwin Hess and Tiit Tammaru (eds), *Housing Estates in the Baltic Countries: The Legacy of Central Planning in Estonia, Latvia and Lithuania* (Cham, Springer, 2019).

21 Andrejs Plakans, *The Latvians: A Short History* (Stanford, CA, Hoover Institution Press, 1995).

22 Hiden and Salmon, *The Baltic Nations*.

23 Paula Christie, 'The Baltic Chain: A study of the organisation facets of large-scale protest from a micro-level perspective', *Lithuanian Historical Studies*, vol. 20 (2015), pp. 183–211.

24 Peter Albertins, 'Unofficial Latvian political organizations in the Gorbachev era: Their evolution, accomplishments, and goals', undergraduate thesis, Western Michigan University, Kalamazoo, 1990.

25 Olgerts Eglitis, *Nonviolent Action in the Liberation of Latvia* (Cambridge, MA, Albert Einstein Institution, 1993).

26 Christie, 'The Baltic Chain'.

27 Daina Stukuls, 'From opposition to independence: Social movements in Latvia, 1986–1991', Center for Research on Social Organization Working Paper 569, 1998, https://deepblue.lib.umich.edu/bitstream/handle/2027.42/51333/569.pdf.

28 Laure Neumayer, 'Bridges across the Atlantic? Intertwined anti-communist mobilisations in Europe and the United States after the Cold War', *Revue d'études comparatives Est-Ouest*, vol. 51, nos 2–3 (2020), pp. 151–83.

29 Eglitis, *Nonviolent Action*.

30 Karsten Brüggemann and Andres Kasekamp, '"Singing oneself into a nation"? Estonian song festivals as rituals of political mobilisation', *Nations and Nationalism*, vol. 20, no. 2 (2014), pp. 259–76; Šmidchens, *The Power of Song*.

31 Stukuls, 'From opposition to independence'.

32   Albertins, 'Unofficial Latvian'.

33   Dainis Īvāns, 'Power of solidarity: Unprecedented cooperation between Baltic independence movements', in *The Baltic Way. Continued* (Rīga, National Museum of Latvia, 2019), pp. 21–31, http://lnvm.lv/wp-content/uploads/2020/02/The-Baltic-Way.-Continued.pdf.

34   UNESCO, 'The Baltic Way: Human chain linking three states in their drive for freedom (Estonia, Latvia, Lithuania)', *Memory of the World Register*, https://media.unesco.org/sites/default/files/webform/mow001/baltic_way.pdf.

35   'The Baltic Way: 30 years on from a day that changed history', LSM, 23 August 2019, https://eng.lsm.lv/article/culture/history/the-baltic-way-30-years-on-from-a-day-that-changed-history.a329603.

36   Helen Wright and Silver Tambur, 'The Baltic Way: The longest unbroken human chain in history', *Estonian World*, 23 August 2021, https://estonianworld.com/life/estonia-commemorates-30-years-since-the-baltic-way-the-longest-unbroken-human-chain-in-history.

37   *The Baltic Way 30* (Rīga, State Chancellery of Latvia, 2019), www.mk.gov.lv/en/media/8837/download.

38   Christie, 'The Baltic Chain'.

39   'The Baltic Way', LSM.

40   Benjamin Peters, *How Not to Network a Nation: The Uneasy History of the Soviet Union* (Cambridge, MA, MIT Press, 2016).

41   Wright and Tambur, 'The Baltic Way'.

42   *The Baltic Way 30*.

43   Christie, 'The Baltic Chain'; Steve Wilmer, 'The Baltic Way as a political performance of subjectivization', in Friedemann Kreuder et al. (eds), *Theater und Subjektkonstitution: Theatrale Praktiken zwischen Affirmation und Subversion* (Bielefeld, transcript, 2012), pp. 655–62.

44   Christie, 'The Baltic Chain'.

45   Ministry of the Economy and Innovation of the Republic of Lithuania, *The Baltic Way*, 2020, https://eimin.lrv.lt/uploads/eimin/documents/files/EN_Baltijos_kelias_2020-10-12_WEB(1).pdf.

46   Michael Dobbs, 'Baltic states link in protest "so our children can be free"', *Washington Post*, 24 August 1989.

47   Linas Jegelevicius, 'Meet the pilot who defied KGB orders to drop flowers on Baltic Way human chain', *Euronews*, 23 August 2019, www.euronews.com/2019/08/23/meet-the-pilot-who-defied-kgb-orders-to-drop-flowers-on-baltic-way-human-chain.

48   Dobbs, 'Baltic states link'.

49 *Congressional Record: Senate*, 19 September 1989, p. 20780.

50 Ibid.

51 David Remnick, 'Kremlin condemns Baltic nationalists; Soviets warn separatism risks disaster', *Washington Post*, 27 August 1989; Esther B. Fein, 'Moscow condemns nationalist "virus" in 3 Baltic lands', *New York Times*, 27 August 1989.

52 Remnick, 'Kremlin condemns'.

53 Dobbs, 'Baltic states link'; 'The Baltic Way to independence: 30 years since the landmark human-chain protest', LRT, 23 August 2019, www.lrt.lt/en/news-in-english/19/1090170/the-baltic-way-to-independence-30-years-since-the-landmark-human-chain-protest.

54 Esther B. Fein, 'Soviets confirm Nazi pacts dividing Europe', *New York Times*, 19 August 1989.

55 Remnick, 'Kremlin condemns'; Esther B. Fein, 'Soviet Congress condemns '39 pact that led to annexation of Baltics', *New York Times*, 25 December 1989.

56 'The dark legacy of Hitler and Stalin in the Baltic', *Congressional Record: Senate*, 19 September 1989, pp. 20779–81, www.govinfo.gov/content/pkg/GPO-CRECB-1989-pt15/pdf/GPO-CRECB-1989-pt15-5-2.pdf.

57 Wright and Tambur, 'The Baltic Way'.

58 Una Bergmane, *Politics of Uncertainty: The United States, the Baltic Question, and the Collapse of the Soviet Union* (Oxford, Oxford University Press, 2023).

59 'AKTO Dėl Lietuvos nepriklausomos valstybės atstatymo', Lietuvos Respublikos Aukščiausiosios Tarybos, 11 March 1990, www.lrs.lt/datos/kovo11/signatarai; Masha Hamilton, 'Lithuania move to secede called "illegal, invalid"', *Los Angeles Times*, 14 March 1990.

60 David Lektzian and Rimvydas Ragauskas, 'The great blockade of Lithuania: Evaluating sanction theory with a case study of Soviet sanctions to prevent Lithuanian independence', *International Area Studies Review*, vol. 19, no. 4 (2016), pp. 320–39.

61 Robertas Povilaitis et al., 'The events of January 1991: Coping with the traumatic experience of independence defenders and other victims', in Danutė Gailienė (ed.), *Lithuanian Faces after Transition: Psychological Consequences of Cultural Trauma* (Vilnius, Eugrimas, 2015), pp. 166–96.

62 *Glasnost in Jeopardy: Human Rights in the USSR* (New York, Helsinki Watch, 1991).

63 Dainis Īvāns, 'Barricades in Latvia', Konrad-Adenauer-Stiftung, 2021,

www.kas.de/documents/252038/10987758/January+1991.+Barricades+in
+Latvia.pdf/ad8a254f-6fb7-0722-6164-e8c94af57544?version=1.0&t=
1610631421871.

64 Eglitis, *Nonviolent Action*; Edgars Engīzers, 'The assessment of the
period of the Barricades by young scholars', www.academia.edu/
31007708/The_historical_significance_of_the_Barricades_in_Riga_
January_1991_.

65 Īvāns, 'Barricades in Latvia'.

66 'Five stories from the 1991 Rīga barricades', LSM, 20 January 2020,
https://eng.lsm.lv/article/culture/history/five-stories-from-the-1991-riga-
barricades.a345406.

67 Eglitis, *Nonviolent Action*.

68 Elizabeth Shogren, 'Over 100,000 join protest against Gorbachev in
Moscow: Dissent: Crowd assails nation's leaders as turning reactionary.
Sympathy is expressed for Lithuanians slain by Soviet troops', *Los
Angeles Times*, 21 January 1991.

69 'The Commemorative Medal for Participants of the Barricades of
1991', Latvijas Republikas Saeima, www.saeima.lv/en/about-saeima/
the-commemorative-medal-for-participants-of-the-barricades.

70 'Various events to commemorate the Barricades 1991 will be held in
Rīga', Rīga City Council, 10 January 2023, www.riga.lv/en/article/
various-events-commemorate-barricades-1991-will-be-held-riga?utm_
source=https%3A%2F%2Fwww.google.com%2F.

71 *Human Rights Watch World Report 1992* (New York, Human Rights
Watch, 1992); Ainius Lasas, 'Bloody Sunday: What did Gorbachev know
about the January 1991 events in Vilnius and Riga?', *Journal of Baltic
Studies*, vol. 38, no. 2 (2007), pp. 179–94.

72 Bergmane, *Politics of Uncertainty*; 'Soviets recall envoy after Iceland
recognizes Lithuania', *Los Angeles Times*, 14 February 1991.

73 Nils R. Muiznieks, 'The influence of the Baltic popular movements on
the process of Soviet disintegration', *Europe-Asia Studies*, vol. 47, no. 1
(1995), pp. 3–25.

74 Bergmane, *Politics of Uncertainty*.

75 Muiznieks, 'The influence'.

76 Daiva Venckus, 'In Vilnius during the 1991 August Moscow coup and
the deadly KGB attack', 19 August 2018, https://daivavenckus.com/
in-vilnius-during-the-1991-august-moscow-coup.

77 Bergmane, *Politics of Uncertainty*.

78 John B. Dunlop, 'The August 1991 coup and its impact on Soviet politics', *Journal of Cold War Studies*, vol. 5, no. 1 (2003), pp. 94–127.

79 Thom Shanker, 'Ukraine votes for declaration of independence', *Chicago Tribune*, 25 August 1991.

80 Bergmane, *Politics of Uncertainty*.

81 'Countries and territories', Freedom House, 2024, https://freedomhouse. org/countries/freedom-world/scores?sort=desc&order=Total%20 Score%20and%20Status.

82 *Global Wealth Databook 2022: Leading Perspectives to Navigate the Future* (Credit Suisse Research Institute, 2022); World Bank, 'Ease of doing business rank', 2019. https://data.worldbank.org/indicator/IC.BUS. EASE.XQ?most_recent_value_desc=false.

83 Daina S. Eglitis and Laura Ardava, 'The politics of memory: Remembering the Baltic Way 20 years after 1989', *Europe-Asia Studies*, vol. 64, no. 6 (2012), pp. 1033–59.

84 Agnese Lāce, 'Russian-speakers in Latvia', Friedrich-Ebert-Stiftung, 2023, https://baltic.fes.de/fileadmin/user_upload/Russian-speakers-in-Latvia.pdf; David J. Trimbach, '"Nationality is ethnicity": Estonia's problematic citizenship policy', Foreign Policy Research Institute, 7 March 2017, www.fpri.org/article/2017/03/nationality-ethnicity-estonias-problematic-citizenship-policy.

85 Una Bergmane, 'How Putin is rehabilitating the Nazi-Soviet pact', Foreign Policy Research Institute, 28 July 2020, www.fpri.org/article/2020/07/putin-rehab-nazi-soviet-pact.

86 'Bilateral aid allocations to Ukraine as a share of 2021 donor country gross domestic product (GDP) between January 24, 2022 and June 30, 2024, by country', *Statista*, 2024, www.statista.com/statistics/1303450/bilateral-aid-to-ukraine-in-a-percent-of-donor-gdp.

87 Joint Communiqué of the Baltic Defence Cooperation Ministerial Committee, 21 December 2021, https://kaitseministeerium.ee/sites/default/files/3B/3b_mc_joint_communique_21_december_2021.pdf.

88 'Riga vice mayor calls for nationalization of Moscow House', *Baltic Times*, 23 February 2023, www.baltictimes.com/riga_vice_mayor_calls_for_nationalization_of_moscow_house; 'Sale of Rīga's "Moscow House" will raise funds for Kyiv', LSM, 30 May 2024, https://eng.lsm.lv/article/politics/saeima/30.05.2024-sale-of-rigas-moscow-house-will-raise-funds-for-kyiv.a556023.

89 John R. Deni, 'Is the Baltic Sea a NATO lake?', Carnegie Endowment for International Peace, 18 December 2023, https://carnegieendowment.org/

research/2023/12/is-the-baltic-sea-a-nato-lake?lang=en; Helga Kalm, 'NATO's path to securing undersea infrastructure in the Baltic Sea', Carnegie Endowment for International Peace, 29 May 2024, https://carnegieendowment.org/research/2024/05/nato-baltic-sea-security-nord-stream-balticconnector?lang=en.

90 Wilmer, 'The Baltic Way'.

91 Guntis Šmidchens, 'A chain of friendship: Reflections on the Baltic Way and inspiration for Belarus', Foreign Policy Research Institute, 31 August 2020, www.fpri.org/article/2020/08/a-chain-of-friendship-reflections-on-the-baltic-way-and-inspiration-for-belarus.

92 'Human chain protest spans Taiwan', CNN, 28 February 2004, https://edition.cnn.com/2004/WORLD/asiapcf/02/28/taiwan.protest.reut.

93 Paul Goldman and Linda Givetash, 'Hong Kong protesters form 28-mile human chain demanding democracy', NBC News, 23 August 2019, www.nbcnews.com/news/world/hong-kong-protesters-form-28-mile-human-chain-demanding-democracy-n1045716.

94 Stephen Burgen and Paul Hamilos, 'Catalans join hands in huge human chain for independence from Spain', *Guardian*, 11 September 2013, www.theguardian.com/world/2013/sep/11/catalans-join-hands-human-chain-independence-spain.

95 Ella Karapetyan, 'Baltics commemorate with relay race', *Baltic Times*, 19 August 2009, www.baltictimes.com/news/articles/23363.

96 '"Baltic Way 30": The first day of a vintage car tour', Antique Automobile Club 'Klasika', 19 August 2019, www.klasika.us/en/2019/08/18/pirmoji-diena; 'The 30th anniversary of The Baltic Way will be marked with a series of significant events throughout Latvia', Ministry of Culture, Republic of Latvia, 1 August 2019, www.km.gov.lv/en/article/30th-anniversary-baltic-way-will-be-marked-series-significant-events-throughout-latvia.

97 'Baltic way anniversary marked with human chain, car drive', ERR News, 24 August 2021.

98 'Thousands form human chain in Lithuania in solidarity with Belarus', Euronews, 24 August 2020, www.euronews.com/2020/08/24/thousands-form-human-chain-in-lithuania-in-solidarity-with-belarus.

6. Restoration

1 Quoted in Jonathan Watts, 'Africa's Great Green Wall just 4% complete halfway through schedule', *Guardian*, 7 September

2020, www.theguardian.com/environment/2020/sep/07/africa-great-green-wall-just-4-complete-over-halfway-through-schedule.

2   Elvis Paul Tangem and Elvis Lyonga Edimo, 'How a Great Green Wall could help Africa combat climate change', World Economic Forum, 18 June 2021, www.weforum.org/agenda/2021/06/how-a-great-green-wall-could-help-africa-combat-climate-change.

3   'Challenges', Great Green Wall, https://thegreatgreenwall.org/challenges.

4   Matthew D. Turner et al., 'Environmental rehabilitation and the vulnerability of the poor: The case of the Great Green Wall', *Land Use Policy*, vol. 11 (2021), 105750; Natalie Thomas and Sumant Nigam, 'Twentieth-century climate change over Africa: Seasonal hydroclimate trends and Sahara Desert expansion', *Journal of Climate*, vol. 31, no. 9 (2018), pp. 3349–70.

5   'Sahel Adaptive Social Protection Program', World Bank Group, www.worldbank.org/en/programs/sahel-adaptive-social-protection-program-trust-fund/overview; Malcolm Potts et al., *Crisis in the Sahel: Possible Solutions and the Consequences of Inaction*, 2013, https://nature.berkeley.edu/release/oasis_monograph_final.pdf.

6   Paul-Arthur Monerie, Benjamin Pohl and Marco Gaetani, 'The fast response of Sahel precipitation to climate change allows effective mitigation action', *npj Climate and Atmospheric Science*, vol. 4 (2021), 24; Mukhtar Balarabe, Khiruddin Abdullah and Mohd Nawawi, 'Long-term trend and seasonal variability of horizontal visibility in Nigerian troposphere', *Atmosphere*, vol. 6, no. 10 (2015), pp. 1462–86.

7   Jill Filipovic, 'Why have four children when you could have seven? Family planning in Niger', *Guardian*, 15 March 2017.

8   'Fertility rate, total (births per woman)', World Bank, 2022, https://data.worldbank.org/indicator/SP.DYN.TFRT.IN?most_recent_value_desc=true.

9   'Sahel Climate, Peace and Security Forum: Climate policy and financing for maintaining peace and security', United Nations Development Programme, 2023, www.undp.org/sites/g/files/zskgke326/files/2023-11/concept-note-forume-climat-en-2.pdf.

10   'Demographic challenges of the Sahel', Population Reference Bureau, 2015, www.prb.org/resources/demographic-challenges-of-the-sahel/.

11   Aryn Baker, 'Can a 4,815-mile wall of trees help curb climate change in Africa', *TIME*, 12 September 2019, https://time.com/5669033/great-green-wall-africa.

12   'Millions face harm from flooding across West and Central Africa,

UNHCR warns', UNHCR, 28 October 2022, www.unhcr.org/us/
news/briefing-notes/millions-face-harm-flooding-across-west-and-
central-africa-unhcr-warns; '"Locust swarm" forces Ethiopian Airlines
plane to divert', BBC News, 13 January 2020, www.bbc.com/news/
world-africa-51098209.

13 Chris Reij, Gray Tappan and Melinda Smale, *Agroenvironmental
Transformation in the Sahel: Another Kind of "Green Revolution"*,
International Food Policy Research Institute (IFPRI) Discussion Paper
00914, 2009.

14 Ellis Adjei Adams et al., 'Farmer–herder conflicts in sub-Saharan
Africa: Drivers, impacts, and resolution and peacebuilding strategies',
Environmental Research Letters, vol. 18 (2023), 123001.

15 *Mali's Young 'Jihadists': Fuelled by Faith or Circumstance?*, Institute
for Security Studies (ISS) policy brief 89, 2016, https://issafrica.
s3.amazonaws.com/site/uploads/policybrief89-eng-v3.pdf; Ahmadou Aly
Mbaye and Landry Signé, *Climate Change, Development, and Conflict-
Fragility Nexus in the Sahel*, Brookings Global Working Paper 169, 2022,
www.brookings.edu/wp-content/uploads/2022/03/Climate-development-
Sahel_Final.pdf.

16 'Burkina Faso: Drone strikes on civilians apparent war crimes', Human
Rights Watch, 25 January 2024, www.hrw.org/news/2024/01/25/
burkina-faso-drone-strikes-civilians-apparent-war-crimes.

17 Oli Brown and Janani Vivekananda, 'Lake Chad shrinking? It's a story
that masks serious failures of governance', *Guardian*, 22 October 2019.

18 'Burkina Faso: Over a quarter million people victims of new "water
war" in peak dry season', Norwegian Refugee Council (NRC), 3 May
2022, www.nrc.no/news/2022/may/burkina-faso-over-a-quarter-million-
people-victims-of-new-water-war-in-peak-dry-season; 'Burkina Faso:
Violence against health care in conflict, 2021', Safeguarding Health
in Conflict Coalition, 2021, https://insecurityinsight.org/wp-content/
uploads/2022/05/2021-Burkina-Faso-SHCC-Factsheet.pdf.

19 'Niger: After two years of activities, MSF teams forced to leave Maïné
Soroa, in Diffa region', Médecins Sans Frontières, 9 August 2019,
www.doctorswithoutborders.org/latest/niger-after-two-years-activities-
msf-teams-forced-leave-maine-soroa-diffa-region.

20 'The Sahel: A deadly new era in the decades-long conflict', Armed
Conflict Location and Event Data (ACLED), 17 January 2024,
https://acleddata.com/conflict-watchlist-2024/sahel; 'Sahel situation',

UNHCR, https://reporting.unhcr.org/operational/situations/sahel-situation.

21 'LDCs at a glance', United Nations, www.un.org/development/desa/dpad/least-developed-country-category/ldcs-at-a-glance.html.

22 Veerle Linseele, 'From the first stock keepers to specialised pastoralists in the West African savannah', in Michael Bollig, Michael Schnegg and Hans-Peter Wotzka (eds), *Pastoralism in Africa: Past, Present and Future* (New York, Berghahn, 2013), pp. 145–70.

23 Eric Ross, 'A historical geography of the Trans-Saharan trade', in Graziano Krätli and Ghislaine Lydon (eds), *The Trans-Saharan Book Trade: Manuscript Culture, Arabic Literacy and Intellectual History in Muslim Africa* (Leiden, Brill, 2011), pp. 1–34.

24 https://thegreatgreenwall.org.

25 'Nearly one million children under 5 in the central Sahel facing severe wasting in 2023 – UNICEF', 7 April 2023, www.unicef.org/wca/press-releases/nearly-one-million-children-under-5-central-sahel-facing-severe-wasting-2023-unicef.

26 'Great Green Wall Initiative', United Nations Convention to Combat Desertification (UNCCD), www.unccd.int/our-work/ggwi.

27 Gregory Mann, 'French colonialism and the making of the modern Sahel', in Leonardo A Villalón (ed.), *The Oxford Handbook of the African Sahel* (Oxford, Oxford University Press, 2021), pp. 35–50.

28 Benjamin E. Thomas, 'Railways and ports in French West Africa', *Economic Geography*, vol. 33, no. 1 (1957), pp. 1–15.

29 Margaret O. McLane, 'Railways or waterways: The Dakar railway network and the Senegal River in the *mise en valeur* of French West Africa', *Proceedings of the Meeting of the French Colonial Historical Society*, vol. 16 (1992), pp. 97–114.

30 Mawa Karambiri et al., '"Trees are not all the same": Assessing the policy and regulatory barriers to the upscaling of Famer Managed Natural Regeneration (FMNR) in Senegal', *Forests, Trees and Livelihoods*, vol. 32, no. 4 (2023), pp. 221–43.

31 Jean B. Faye and Yvonne A. Braun, 'Soil and human health: Understanding agricultural and socio-environmental risk and resilience in the age of climate change', *Health & Place*, vol. 77 (2022), 102799.

32 Jim Morrison, 'The "Great Green Wall" didn't stop desertification, but it evolved into something that might', *Smithsonian Magazine*, 23 August 2016, www.smithsonianmag.com/science-nature/great-green-wall-stop-desertification-not-so-much-180960171.

33 Karambiri et al., '"Trees are not"'.

34 Nicholas Atampugre, *Behind the Lines of Stone: The Social Impact of a Soil and Water Conservation Project in the Sahel* (Oxford, Oxfam, 1993).

35 Reij et al., *Agroenvironmental Transformation*.

36 Ibid.

37 Jan Sendzimir, Chris P. Reij and Piotr Magnuszewski, 'Rebuilding resilience in the Sahel: Regreening in the Maradi and Zinder regions of Niger', *Ecology and Society*, vol. 16, no. 3 (2011).

38 Atampugre, *Behind the Lines*.

39 Matthew D. Turner et al., 'Great Green Walls: Hype, myth, and science', *Annual Review of Environment and Resources*, vol. 48 (2023), pp. 263–87.

40 Marie Ladekjær Gravesen and Mikkel Funder, *An Overview and Lessons Learnt: The Great Green Wall* (Copenhagen, Danish Institute for International Studies (DIIS), 2022).

41 Natalie R. Wilkinson, 'The Great Green Wall: A continuance of Sahelian adaptation', *Environmental History Now (EHN)*, 14 July 2021, https://envhistnow.com/2021/07/14/the-great-green-wall-a-continuance-of-sahelian-adaptation.

42 Anna Alsobrook, 'The social impacts of the Great Green Wall in rural, Senegalese villages', *Journal of Sustainable Development in Africa*, vol. 17, no. 2 (2015), pp. 130–49.

43 Morrison, 'The "Great Green Wall" didn't stop desertification'.

44 Turner et al., 'Great Green Walls'.

45 'Dead baby trees by the millions as reforestation fails', *New Humanitarian*, 8 April 2008, www.thenewhumanitarian.org/news/2008/04/08/dead-baby-trees-millions-reforestation-fails.

46 Alsobrook, 'The social impacts'.

47 Cheikh Mbow, 'The Great Green Wall in the Sahel', *Oxford Research Encyclopedia of Climate Science*, 2017, https://doi.org/10.1093/acrefore/9780190228620.013.559.

48 Reij et al., *Agroenvironmental Transformation*; Raphael Belmin, Hamado Sawadogo and Moussa N'Dienor, 'The zaï technique: How farmers in the Sahel grow crops with little to no water', *Conversation*, 27 December 2023, https://theconversation.com/the-za-technique-how-farmers-in-the-sahel-grow-crops-with-little-to-no-water-220103.

49 Will Critchley and Klaus Siegert, *A Manual for the Design and Construction of Water Harvesting Schemes for Plant Production* (Rome, Food and Agriculture Organization of the United Nations, 1991).

50 Reij et al., *Agroenvironmental Transformation*.

51 Jenny C. Aker and B. Kelsey Jack, *Harvesting the Rain: The Adoption of Environmental Technologies in the Sahel*, National Bureau of Economic Research (NBER) Working Paper No. 29518, 2021, www.nber.org/system/files/working_papers/w29518/w29518.pdf.

52 Johan A. van Dijk, 'Indigenous and introduced: Soil and water conservation in Sudan', *Waterlines*, vol. 13, no. 4 (1995), pp. 19–21; *The Great Green Wall Implementation Status*, UNCCD.

53 Lindsay Cobb, '1 million trees planted in Chad', *Trees for the Future (TREES)*, 26 July 2021, https://trees.org/2021/07/26/one-million-chad.

54 Hazel Healy, 'Taking back the peanut basin', *New Internationalist*, 20 September 2021, https://newint.org/features/2021/08/09/land-beautiful-agroecology-senegal-food-fjf.

55 'Voices: Tony Rinaudo', Reforestation World, www.reforestationworld.org/voices-tony-rinaudo-principal-advisor-natural-resources-world-vision-australia/.

56 Tony Rinaudo, 'The development of Farmer Managed Natural Regeneration', *LEISA Magazine*, vol. 23, no. 2 (2007), pp. 32–4.

57 Atampugre, *Behind the Lines*.

58 Reij et al., *Agroenvironmental Transformation*.

59 A. C. Franke et al., 'Sustainable intensification through rotations with grain legumes in Sub-Saharan Africa: A review', *Agriculture, Ecosystems & Environment*, vol. 261 (2018), pp. 172–85.

60 'In Burkina Faso, the Great Green Wall is taking shape', Food and Agriculture Organization of the United Nations (FAO), 5 July 2019, www.fao.org/in-action/action-against-desertification/news-and-multimedia/detail/en/c/1200852.

61 Morrison, 'The "Great Green Wall"'.

62 Reij et al., *Agroenvironmental Transformation*.

63 Nicholas E. Young et al., 'Twenty-three years of forest cover change in protected areas under different governance strategies: A case study from Ethiopia's southern highlands', *Land Use Policy*, vol. 91 (2020), 104426; 'Forests, grasslands, and drylands: Ethiopia', World Resources Institute (WRI), 2003, www.idp-uk.org/OurProjects/Environment/Forests,grasslands,%20drylands(FAO)%20%20for_cou_231.pdf.

64 Alison Abbott, 'Biodiversity thrives in Ethiopia's church forests', *Nature*, 29 January 2019, www.nature.com/immersive/d41586-019-00275-x/index.html.

65 *Global Land Outlook: Land Restoration for Recovery and Resilience*, second edition (Bonn, UNCCD, 2022).

66 *The Great Green Wall Implementation Status*, UNCCD.

67 https://greenlegacy.et/green-legacy/home.

68 'Ethiopia plants 566 million trees seedlings in 12 hours', *FurtherAfrica*, 25 July 2023, https://furtherafrica.com/2023/07/25/ethiopia-plants-566-million-trees-seedlings-in-12-hours; 'Ethiopia plants over 350 million trees in a day, setting new world record', United Nations Environment Programme (UNEP), 2 August 2019, www.unep.org/news-and-stories/story/ethiopia-plants-over-350-million-trees-day-setting-new-world-record.

69 'Senegal', Climate and Clean Air Coalition (CCAC), www.ccacoalition.org/partners/senegal.

70 Spoorthy Raman, 'Progress is slow on Africa's Great Green Wall, but some bright spots bloom', *Mongabay*, 3 August 2023, https://news.mongabay.com/2023/08/progress-is-slow-on-africas-great-green-wall-but-some-bright-spots-bloom.

71 'Growing the Olympic Forest', Tree Aid, www.treeaid.org/projects/mali/olympic-forest/.

72 Turner et al., 'Great Green Walls'.

73 *Southern Africa Thematic Report: Leveraging the Land, Water and Energy Nexus in SADC*, United Nations Convention to Combat Desertification (UNCCD), 2022.

74 'Desert to Power', African Development Bank Group, 2021, www.afdb.org/sites/default/files/news_documents/dtp-brochure-2021.pdf.

75 Sohaib Mahmoud and Mohamed Taifouri, 'The coups d'état of the Sahel region: Domestic causes and international competition', Arab Center Washington DC, 27 September 2023, https://arabcenterdc.org/resource/the-coups-detat-of-the-sahel-region-domestic-causes-and-international-competition; 'Independent review of the Great Green Wall Accelerator: Final report', United Nations Convention to Combat Desertification (UNCCD), 2023, www.unccd.int/sites/default/files/inline-files/GGWA%20review%20final%20report%20formatted.pdf.

76 'Burkina Faso: People caught in a perfect storm of conflict, displacement and food insecurity', Internal Displacement Monitoring Centre (IDMC), 5 June 2023, https://story.internal-displacement.org/burkina-faso-People-caught-in-a-perfect-storm-of-conflict-displacement-and-food-insecurity/index.html.

77 'Burkina Faso', UNHCR, 2024, https://reporting.unhcr.org/operational/

operations/burkina-faso; *Aperçu des Besoins Humanitaires*, United Nations Office for the Coordination of Humanitarian Affairs (OCHA), 2023; 'Burkina Faso: Armed groups committing war crimes in besieged localities', Amnesty International, 2 November 2023, www.amnesty.org/en/latest/news/2023/11/202715.

78 *Firearms Trafficking in the Sahel: Transnational Organized Crime Threat Assessment – Sahel*, United Nations Office on Drugs and Crime (UNODC), 2022.

79 *Gold Trafficking in the Sahel: Transnational Organized Crime Threat Assessment – Sahel*, United Nations Office on Drugs and Crime (UNODC), 2022.

80 Jessica Berlin et al., *The Blood Gold Report: How the Kremlin Is Using Wagner to Launder Billions in African Gold* (Democracy 21, Washington, DC, 2023), https://bloodgoldreport.com/wp-content/uploads/2023/12/The-Blood-Gold-Report-2023-December.pdf.

81 Simeon Ehui, Chakib Jenane and Kaja Waldmmann, 'The war in Ukraine: Amplifying an already prevailing food crisis in West Africa and the Sahel region', *World Bank Voices*, 13 April 2022, https://blogs.worldbank.org/en/voices/war-ukraine-amplifying-already-prevailing-food-crisis-west-africa-and-sahel-region; Robert E. Clute, 'The role of agriculture in African development', *African Studies Review*, vol. 25, no. 4 (1982), pp. 1–20.

82 Daniel Muteti, 'Sudan crisis spells uncertainty for key Coca-Cola ingredient',DW,20July2023,www.dw.com/en/sudan-crisis-spells-uncertainty-for-key-coca-cola-ingredient/a-66256836.

83 Jessica Tyler, 'There's a church in Mexico where Coca-Cola is used in religious ceremonies', *Business Insider*, 22 August 2018, www.businessinsider.com/coca-cola-church-in-mexico-uses-coke-religious-ceremonies-2018-8.

84 Oscar Lopez and Andrew Jacobs, 'In town with little water, Coca-Cola is everywhere. So is diabetes', *New York Times*, 14 July 2018.

85 'Continental priorities', African Union, https://au.int/en/aureforms/priorities.

86 'Flagship projects of Agenda 2063', African Union, https://au.int/agenda2063/flagship-projects.

87 *The Great Green Wall Implementation Status*, UNCCD, 2020, p. 36.

88 Tristan Bove, 'The Great Green Wall is failing, but its legacy could still be a success', Earth.org, 24 March 2021, https://earth.org/the-great-green-wall-legacy; Daniel Cusick, 'Will Africa ever see its "Great Green

Wall"?', E&E News, 4 November 2022, www.eenews.net/articles/
will-africa-ever-see-its-great-green-wall.

89 'Independent review', UNCCD.

90 'How to make Africa's "Great Green Wall" a success', *Nature*, 4 May
2022, www.nature.com/articles/d41586-022-01201-4.

91 'Independent review', UNCCD; Gravesen and Funder, *The Great Green
Wall*.

92 Turner et al., 'Environmental rehabilitation'.

93 *Independent Review*, UNCCD; *The Great Green Wall Initiative:
Supporting Resilient Livelihoods and Landscapes in the Sahel*, Global
Environment Facility (GEF), 2023.

94 Karambiri et al., '"Trees are not"'.

95 Bove, 'The Great Green Wall'; Anthony O. Onoja and Anthonia I.
Achike, 'Large-scale land acquisitions by foreign investors in West Africa:
Learning points', *Consilience*, vol. 14 (2015), pp. 173–88.

96 Karambiri et al., '"Trees are not"'.

97 Kanyinke Sena, 'Recognizing Indigenous peoples' land interests is critical
for people and nature', World Wildlife Fund (WWF), 22 October 2020,
www.worldwildlife.org/stories/recognizing-indigenous-peoples-land-
interests-is-critical-for-people-and-nature.

98 Food and Agriculture Organization of the United Nations (FAO) and
United Nations Environment Programme (UNEP), *The State of the
World's Forests 2020: Forests, Biodiversity and People* (Rome, FAO
and UNEP, 2020), www.fao.org/3/ca8642en/online/ca8642en.html;
Independent Review, UNCCD.

99 *Independent Review*, UNCCD; Raman, 'Progress is slow'.

100 'Green Wall Accelerator', United Nations Convention to Combat
Desertification (UNCCD), www.unccd.int/our-work/ggwi/
great-green-wall-accelerator.

101 'New observatory to track progress of Africa's Great Green Wall',
UNCCD, 3 June 2024, www.unccd.int/news-stories/statements/
new-observatory-track-progress-africas-great-green-wall.

102 *The Great Green Wall Implementation Status*, UNCCD; Gravesen and
Funder, *The Great Green Wall*.

103 Amadou Ndiaye, 'Practices of the Great Green Wall project in the Ferlo
(Senegal): Effects on pastoral resilience and development', *World Journal
of Science*, vol. 3, no. 2 (2016).

104 Alsobrook, 'The social impacts'.

105 Turner et al., 'Environmental rehabilitation'; Ndiaye, 'Practices of the Great Green Wall'.

106 Turner et al., 'Environmental rehabilitation'.

107 Moctar Sacande et al., 'Socio-economic impacts derived from large scale restoration in three Great Green Wall countries', *Journal of Rural Studies*, vol. 87 (2021), pp. 160–81.

108 Abdullahi Mohammad Jalam et al., 'Deployment of performance indicators toward bridging monitoring gaps in Africa's Great Green Wall', *Environmental Health Engineering and Management Journal*, vol. 10, no. 4 (2023), pp. 429–39; 'In Burkina Faso, the Great Green Wall is taking shape', Food and Agriculture Organization of the United Nations (FAO), 5 July 2019, www.fao.org/in-action/ action-against-desertification/news-and-multimedia/detail/en/c/1200852.

109 Alsobrook, 'The social impacts'; UNCCD, *Global Land Outlook*.

110 Pauline Castaing and Antoine Leblois, 'The Great Green Wall, a bulwark against food insecurity? Evidence from Nigeria', Center for Environmental Economics – Montpellier, Working Paper 2023-01, https://hal.inrae.fr/hal-03958274v3/document.

111 'Statement by President von der Leyen on the Great Green Wall at COP26', European Commission, 1 November 2021, https://ec.europa.eu/commission/presscorner/detail/en/ statement_21_5742; 'The Great Green Wall Accelerator: New impetus for this iconic African initiative', Ministère de l'Europe et des Affaires Étrangères, 2022, www.diplomatie.gouv.fr/en/country-files/africa/ climate-issues/article/the-great-green-wall-accelerator-new-impetus-for- this-iconic-african-initiative.

7. Co-option

1 Quoted in W. F. Peck, 'Black Hawk: The man – the hero – the patriot', *Annals of Iowa*, vol. 2, no. 6 (1896), pp. 450–64, at p. 458.

2 Jonathan Boyd, '"Windy City"', *Encyclopedia of Chicago*, 2005, www.encyclopedia.chicagohistory.org/pages/6.html.

3 Carl Sandburg, 'Chicago', *Poetry*, vol. 3, no. 6 (1914).

4 Karen Zraick, 'What is Labor Day? A history of the workers' holiday', *New York Times*, 4 September 2023.

5 Dena Evelyn Shapiro, 'Indian tribes and trails of the Chicago region: A preliminary study of the influence of the Indian on early white settlement', master's thesis, University of Chicago, 1929.

6   Tanner Howard, 'Native American routes: The ancient trails hidden in Chicago's grid system', *Guardian*, 17 January 2019.

7   David Edmunds, 'Chicago in the middle ground', *Encyclopedia of Chicago*, 2005, www.encyclopedia.chicagohistory.org/pages/254.html.

8   Andrew Herscher, 'Settler colonialism in Chicago: A living atlas', *Rampant*, 10 October 2022, https://rampantmag.com/2022/10/settler-colonialism-in-chicago-a-living-atlas.

9   Robert G. Spinney, *City of Big Shoulders: A History of Chicago*, second edition (Ithaca, NY, Cornell University Press, 2020).

10 Carol Flynn, 'Native Americans find balance with nature in local region', *Beverly Review*, 6 December 2022.

11 Edmunds, 'Chicago in the middle ground'; M. J. Morgan, *Land of Big Rivers: French and Indian Illinois, 1699–1778* (Carbondale, Southern Illinois University Press, 2010).

12 John William Nelson, *Muddy Ground: Native Peoples, Chicago's Portage, and the Transformation of a Continent* (Chapel Hill, University of North Carolina Press, 2023).

13 Benjamin Sells, *A History of the Chicago Portage: The Crossroads That Made Chicago and Helped Make America* (Evanston, IL, Northwestern University Press, 2021).

14 Jesse Dukes, 'Without Native Americans, would we have Chicago as we know it?', WBEZ, 12 November 2017.

15 Libby Hill, *The Chicago River: A Natural and Unnatural History*, revised edition (Carbondale, Southern Illinois University Press, 2019).

16 Jacques Marquette, *The Mississippi Voyage of Jolliet and Marquette*, 1673 (repr. Madison, Wisconsin Historical Society, 2003), www.americanjourneys.org/AJ_PDF/AJ-051.pdf.

17 Spinney, *City of Big Shoulders*.

18 Hill, *The Chicago River*.

19 Marquette, *The Mississippi Voyage*.

20 Spinney, *City of Big Shoulders*.

21 Reginald Horsman, 'The Northwest Ordinance and the shaping of an expanding republic', *Wisconsin Magazine of History*, vol. 73, no. 1 (1989), pp. 21–32.

22 Howard, 'Native American routes'.

23 'Thompson's plat of 1830', *Encyclopedia of Chicago*, www.encyclopedia.chicagohistory.org/pages/11175.html.

24 Lawrence B. A. Hatter, *Citizens of Convenience: The Imperial Origins*

of American Nationhood on the US–Canadian Border (Charlottesville, University of Virginia Press, 2017).

25 Ann Durkin Keating, *Rising Up from Indian Country: The Battle of Fort Dearborn and the Birth of Chicago* (Chicago, University of Chicago Press, 2012).

26 Joe Ward, 'A Rogers Park plaque marks how US grabbed Indigenous people's land that became Chicago. Neighbors are deciding what to do with it', *Block Club Chicago*, 6 May 2021; David E. Missirian, 'Native Nations' land ownership and our disservice to their people and culture: A proposed legislative solution and a lesson to be learned', *American Indian Law Journal*, vol. 9, no. 2 (2021), pp. 383–401.

27 Article 1 of Treaty with the Ottawa, Etc., 1816, https://treaties.okstate.edu/treaties/treaty-with-the-ottawa-etc-1816-0132.

28 Keating, *Rising Up*.

29 Spinney, *City of Big Shoulders*.

30 Michael Golay, *The Tide of Empire: America's March to the Pacific* (New York, John Wiley & Sons, 2003).

31 'A clash of cultures', Smithsonian American Art Museum, https://americanexperience.si.edu/historical-eras/expansion/pair-pigeons-egg-head-speculator.

32 Spinney, *City of Big Shoulders*.

33 'Indian treaties and the Removal Act of 1830', Office of the Historian, https://history.state.gov/milestones/1830-1860/indian-treaties.

34 Spinney, *City of Big Shoulders*.

35 Patrick T. Reardon, 'Chicago's trail of tears: Potawatomi warriors' 1835 dance marked eviction', *Chicago Tribune*, 24 August 2016.

36 Walter Nugent, 'Demography: Chicago as a modern world city', *Encyclopedia of Chicago*, 2005, https://encyclopedia.chicagohistory.org/pages/962.html.

37 Sells, *A History of the Chicago Portage*.

38 Hill, *The Chicago River*.

39 William Cronon, *Nature's Metropolis: Chicago and the Great West* (New York, W. W. Norton, 1991).

40 Dominic A. Pacyga, *Slaughterhouse: Chicago's Union Stock Yard and the World It Made* (Chicago, University of Chicago Press, 2015).

41 David A. Belden, *Illinois and Michigan Canal* (Charleston, SC, Arcadia, 2012).

42 John F. Hogan and Alex A. Burkholder, *A Fire Strikes the Chicago Stock*

Yards: A History of Flame and Folly in the Jungle (Charleston, SC, History Press, 2013).

43 Upton Sinclair, *The Jungle* (New York, Grosset & Dunlap, 1906), p. 112.

44 Gerald W. Adelmann, 'Reworking the landscape, Chicago style', *Hastings Center Report*, vol. 28, no. 6 (1998), pp. S6–S11.

45 'A history of protecting our water environment', Metropolitan Water Reclamation District of Greater Chicago, 2023, https://mwrd.org/what-we-do/history-protecting-our-water-environment.

46 Missouri *v.* Illinois, 200 U.S. 496 (1906).

47 Geoffrey Baer, 'The history of the Chicago River', WTTW, https://interactive.wttw.com/chicago-river-tour/history-chicago-river.

48 Shapiro, *Indian Tribes and Trails*.

49 Keating, *Rising Up*.

50 Edmunds, 'Chicago in the middle ground'.

51 John D. Haeger, 'The American Fur Company and the Chicago of 1812–1835', *Journal of the Illinois State Historical Society*, vol. 61, no. 2 (1968), pp. 117–39.

52 Joel Greenberg, *A Natural History of the Chicago Region* (Chicago, University of Chicago Press, 2002); Dukes, 'Without Native Americans'.

53 Cronon, *Nature's Metropolis*.

54 David Young, 'Wild West Chicago', *Chicago Tribune*, 26 October 1997.

55 Cronon, *Nature's Metropolis*.

56 John C. Hudson,' Railroads', *Encyclopedia of Chicago*, 2005, http://www.encyclopedia.chicagohistory.org/pages/1039.html.

57 Cronon, *Nature's Metropolis*.

58 Jim Sulski, 'Showing the way', *Chicago Tribune*, 18 December 1997.

59 Chicagology, 'State Street', https://chicagology.com/chicagostreets/statestreet.

60 Neil Gale, 'Plank road history in the Chicago area', *Digital Research Library of Illinois History Journal*, 20 November 2019.

61 Daniel H. Burnham and Edward H. Bennett, *Plan of Chicago*, ed. Charles Moore (Chicago, Commercial Club, 1909).

62 'Chicago surface lines: History', Chicago Transit & Railfan, https://chicagorailfan.com/ctabhist.html.

63 Austin Whittall, 'Route 66 from Chicago to Pontiac', TheRoute-66.com, 1 May 2024, www.theroute-66.com/chicago-to-pontiac.html.

64 Howard, 'Native American routes'.

65 Albert F. Scharf, 'Indian trails and villages of Chicago and of Cook, DuPage and Will counties, Ills. (1804) as shown by weapons and

implements of the Stone-Age', 1900/1, https://news.wttw.com/sites/default/files/article/file-attachments/The%20Scharf%20Map.pdf.

66 Greenberg, *A Natural History*.

67 Megan Bang et al., 'Muskrat theories, tobacco in the streets, and living Chicago as Indigenous land', *Environmental Education Research*, vol. 20, no. 1 (2014), pp. 37–55.

68 'What does the word "Chicago" mean?', *Chicagology*, https://chicagology.com/prefire/prefire282.

69 'Origin of names of US states', US Department of the Interior: Indian Affairs, www.bia.gov/as-ia/opa/online-press-release/origin-names-us-states.

70 Sulski, 'Showing the way'.

71 Euan Hague, 'Day 54: A landscape marked by white supremacy', in *Project 100 Days: Challenging White Supremacy*, DePaul University Women's Center/Office of Multicultural Student Success, 2020, pp. 121–2.

72 Kelly Bauer, 'DuSable Lake Shore Drive name change official with new signs honoring city's Black founder', *Block Club Chicago*, 21 October 2023.

73 James B. LaGrand, *Indian Metropolis: Native Americans in Chicago, 1945–75* (Urbana, University of Illinois Press, 2002).

74 C. D. Green, 'A major museum's attempt to center Native American voices', *Sapiens*, 8 December 2022, www.sapiens.org/culture/field-museum-native-voices.

75 https://cpag.squarespace.com/northwest-portage-walking-museum.

76 Christopher Borrelli, 'Artist JeeYeun Lee's walking tours tell a very different story about Chicago and its lakefront', *Chicago Tribune*, 10 August 2023.

77 Daniel Hautzinger, '"We're still here": Chicago's Native American community', WTTW, 8 November 2018.

8. Vitality

1 Quoted in Roger Shepherd, *Baekdu Daegan: Hiking Korea's Mountain Spine* (HIKEKOREA, 2017), p. 301, https://issuu.com/hikekorea/docs/pdf_preview_pages.

2 Cindy Blanco, '2020 Duolingo language report: Global overview', Duolingo blog, 15 December 2020, https://blog.duolingo.com/global-language-report-2020.

3   Choe Sang-Hun, 'Kim Jong-un calls K-pop a "vicious cancer" in the new culture war', *New York Times*, 10 June 2021.

4   Peter Baker and Michael Crowley, 'Trump steps into North Korea and agrees with Kim Jong-un to resume talks', *New York Times*, 30 June 2019.

5   Yong-pyo Hong, *State Security and Regime Security: President Syngman Rhee and the Insecurity Dilemma in South Korea, 1953–60* (New York, St Martin's Press, 2000).

6   Seo-Hyun Park, 'Dueling nationalisms in North and South Korea', *Palgrave Communications*, vol. 5, no. 40 (2019).

7   Allan R. Millett, *The War for Korea, 1945–1950: A House Burning* (Lawrence, University Press of Kansas, 2005).

8   Bruce Cumings, *The Korean War: A History* (New York, Modern Library, 2010).

9   James A. Foley, '"Ten million families": Statistic or metaphor?', *Korean Studies*, vol. 25, no. 1 (2001), pp. 96–110.

10 Steve Lohr, 'War-scattered Korean kin find their kin at last', *New York Times*, 18 August 1983.

11 Roger Shepherd, 'The Baekdu Daegan as a symbol of Korea', Korean Culture and Information Service (KOCIS), 2018, www.kocis.go.kr/eng/webzine/201810/sub02.html.

12 Hong-key Yoon, *The Culture of Fengshui in Korea: An Exploration of East Asian Geomancy* (Lanham, MD, Lexington Books, 2006).

13 David A. Mason, 'Mountain tourism and religious heritage sites: A fresh paradigm', *Transactions of the Royal Asiatic Society Korea Branch*, vol. 87 (2012), pp. 65–80, https://raskb.com/full-texts-by-volume-2.

14 David A. Mason, 'The background and contemporary spiritual-nationalist significance of Mt Baekdu-san and the Baekdu-daegan range, in all of Korea', 2019, https://seed.upol.cz/wp-content/uploads/2019/07/lecture-essay-david-mason.pdf; Susann Valerie Ahn, 'Cultural laboratory Seoul: Emergence, narrative and impact of culturally related landscape meanings', PhD thesis, ETH Zurich, 2019.

15 Hae-Joon Jung, 'Landscape as heritage: Towards a conservation framework for scenic sites in Korea', PhD thesis, University of Sheffield, 2015.

16 Mason, 'The background'.

17 Yoon, *The Culture of Fengshui*.

18 David A. Mason, 'Pungsu-jiri: Korea's system of geomancy or feng shui', www.san-shin.net/Pungsu-jiri.html.

19 Mason, 'Pungsu-jiri'.

20 Mason, 'The background'.

21 Mason, 'Pungsu-jiri'.

22 Sem Vermeersch, 'Buddhism as a cure for the land', in Robert E. Buswell, Jr (ed.), *Religions of Korea in Practice* (Princeton, NJ, Princeton University Press, 2007), pp. 76–85.

23 Roger Shepherd, 'The Baekdu Daegan (백두대간) as a symbol of unity', *Gwangju News*, 9 May 2018, https://gwangjunewsgic.com/travel/around-korea/the-baekdu-daegan.

24 Jung, 'Landscape as heritage'.

25 Jang Jiyeon, 'Korean geomancy from the tenth through the twentieth centuries: Changes and continuities', *Seoul Journal of Korean Studies*, vol. 30, no. 2 (2017), pp. 101–29.

26 Hong-key Yoon, 'The eight periods in the history of Korean geomancy', in Hong-key Yoon (ed.), *P'ungsu: A Study of Geomancy in Korea* (Albany, State University of New York Press, 2017), pp. 23–60.

27 Hong-key Yoon, 'Government affairs relating to geomancy during the time of premodern Korea', in Yoon (ed.), *P'ungsu*, pp. 71–80.

28 Jang, 'Korean geomancy'.

29 Hong-key Yoon, 'Geomancy and social upheavals in Korea', *European Journal of Geopolitics*, vol. 2 (2014), pp. 5–23.

30 Hong-key Yoon, 'Human modification of Korean landforms for geomantic purposes', *Geographical Review*, vol. 101, no. 2 (2011), pp. 243–60.

31 Bo-Chul Whang and Myung-Woo Lee, 'Landscape ecology planning principles in Korean Feng-Shui, Bi-bo woodlands and ponds', *Landscape and Ecological Engineering*, vol. 2 (2006), pp. 147–62.

32 Jang, 'Korean geomancy'.

33 Yoon, *The Culture of Fengshui*.

34 Jung, 'Landscape as heritage'.

35 Dowon Lee et al., 'Korea's sacred groves: The "maeulsoop"', in Chris Coggins and Bixia Chen (eds), *Sacred Forests of Asia: Spiritual Ecology and the Politics of Nature Conservation* (Abingdon, Routledge, 2022), pp. 136–49.

36 Whang and Lee, 'Landscape ecology'.

37 Mason, 'Pungsu-jiri'.

38 Ben Jackson and Robert Koehler, *Korean Architecture: Breathing with Nature* (Seoul, Seoul Selection, 2012).

39 Oh Sang-Hak, 'The recognition of geomancy by intellectuals during

the Joseon period', *Review of Korean Studies*, vol. 13, no. 1 (2010), pp. 121–47.

40 Jung, 'Landscape as heritage'.

41 David A. Mason, 'What makes a Korean mountain "sacred"?', www.san-shin.org/index4.html.

42 Laura Broadwell, 'Top 10 Korean surnames', Holt International, 26 August 2022, www.holtinternational.org/top-10-korean-surnames.

43 Yeong-Kook Choi, 'Baekdudaegan, the central axis of the Korean peninsular [*sic*]: The path toward management strategies regarding to its concepts', in Sun-Kee Hong et al. (eds), *Ecological Issues in a Changing World: Status, Response and Strategy* (Dordrecht, Kluwer, 2004), pp. 355–83.

44 Shin Ik-cheol, 'Travel to Baekdusan and its significance during the Joseon period', *Review of Korean Studies*, vol. 13, no. 4 (2010), pp. 13–31.

45 Jung, 'Landscape as heritage', p. 150.

46 Shin, 'Travel to Baekdusan'; Jo Yoong-hee, 'Westerners' perceptions of Baekdusan until the nineteenth century: Focusing on materials in English', *Review of Korean Studies*, vol. 13, no. 4 (2010), pp. 133–49.

47 Mason, 'The background'.

48 Choi, 'Baekdudaegan'.

49 Jihee Park and Gerald T. Kyle, 'Identifying place meanings ascribed to the Baekdu-Daegan trail in South Korea', *Journal of Leisure Research*, vol. 52, no. 2 (2021), pp. 180–201; Choi, 'Baekdudaegan'.

50 National Geographic Information Institute of the Ministry of Land, Infrastructure and Transport, 'Classification of topographies and basins', *National Atlas of Korea*, 2019, http://nationalatlas.ngii.go.kr/pages/page_1949.php.

51 Mason, 'Mountain tourism'; Hyemin Na, 'Korean diasporic megachurches in the United States', in Afe Adogame et al. (eds), *Routledge Handbook of Megachurches* (Abingdon, Routledge, 2024), pp. 151–63.

52 Choi, 'Baekdudaegan'.

53 Shin, 'Travel to Baekdusan'.

54 Chi-Young, 'The people'.

55 Northeast Asian History Network, 'Korea–China history awareness', http://contents.nahf.or.kr/english/item/level.do?levelId=iscd_003e_0050_0010.

56 Nianshen Song, *Making Borders in Modern East Asia: The Tumen River Demarcation, 1881–1919* (Cambridge, Cambridge University

Press, 2018); 'The Mukden Incident of 1931 and the Stimson Doctrine', Office of the Historian, https://history.state.gov/milestones/1921-1936/mukden-incident.

57 Zhihua Shen and Yafeng Xia, 'Contested border: A historical investigation into the Sino-Korean border issue, 1950–1964', *Asian Perspective*, vol. 37 (2013), pp. 1–30.

58 Jinwung Kim, *A History of Korea: From 'Land of the Morning Calm' to States in Conflict* (Bloomington, Indiana University Press, 2012).

59 Je-Hun Ryu and Doo-Hee Won, 'The modern production of multiple meanings of the Baekdudaegan mountain system', *Korea Journal*, vol. 53, no. 3 (2013), pp. 103–32.

60 Ari L. Goldman, 'Japan's years as harsh ruler deeply embittered Koreans', *New York Times*, 7 September 1984.

61 Kim, *A History*.

62 'South Korea faces backlash from WWII forced labor victims', DW, 7 March 2023, www.dw.com/en/south-korea-faces-backlash-from-wwii-forced-labor-victims/a-64907859.

63 Soyang Park, 'Speaking with the colonial ghosts and *pungsu* rumour in contemporary South Korea (1990–2006): The *pungsu* (feng shui) invasion story surrounding the demolition of the former Japanese Colonial-General building and iron spikes', *Journal for Cultural Research*, vol. 16, no. 1 (2011), pp. 21–42.

64 Eun-Sung Kim et al., 'Korean traditional beliefs and renewable energy transitions: Pungsu, shamanism, and the local perception of wind turbines', *Energy Research & Social Science*, vol. 46 (2018), pp. 262–73.

65 Yoon, *The Culture of Fengshui*.

66 Zac Thompson, 'This (possibly cursed) presidential palace just opened to the public for the first time', Frommer's, 24 May 2022, www.frommers.com/blogs/passportable/blog_posts/this-possibly-cursed-presidential-palace-just-opened-to-the-public-for-the-first-time.

67 Jennifer Billock, 'How Korea's Demilitarized Zone became an accidental wildlife paradise', *Smithsonian Magazine*, 12 February 2018, www.smithsonianmag.com/travel/wildlife-thrives-dmz-korea-risk-location-180967842.

68 Suki Kim, 'The meaning of Kim Jong Nam's murder', *Atlantic*, 24 February 2017, www.theatlantic.com/international/archive/2017/02/north-korea-kim-jong-il-kim-jong-nam-malaysia/517635.

69 Michael Rundle, 'Kim Jong Il dead: "Nature mourning" North Korean leader reports state media', *Huffington Post UK*, 22 December 2011,

www.huffingtonpost.co.uk/2011/12/22/kim-jong-il-dead-nature-mourns_n_1164845.html; 'Kim Jong-il death: "Nature mourns" N Korea leader', BBC News, 22 December 2011, www.bbc.com/news/world-asia-16297811.

70   Joel Lee, 'Trekking on sacred Paektusan highlands', *Korea Herald*, 20 September 2018, www.koreaherald.com/view.php?ud=20180920000786.

71   Kwangbaek Lee and Rose Adams, 'Kim Jong Un abandoned unification. What do North Koreans think?', *Diplomat*, 2 September 2024, https://thediplomat.com/2024/09/kim-jong-un-abandoned-unification-what-do-north-koreans-think.

72   Justin McCurry, 'Korean peninsula will be united by 2045, says Seoul amid Japan row', *Guardian*, 15 August 2019, www.theguardian.com/world/2019/aug/15/korean-peninsula-will-be-united-by-2045-says-seoul-amid-japan-row.

73   David Choi, 'A symbolic tree was planted on the North–South Korean border during the groundbreaking summit', *Business Insider*, 27 April 2018, www.businessinsider.com/north-and-south-korea-tree-planting-dmz-2018-4.

74   Seong Yeon-cheol, 'Moon and Kim ascend Mount Baekdu together', *Hankyoreh*, 29 September 2018, https://english.hani.co.kr/arti/english_edition/e_northkorea/862987.

75   Quoted in ibid.

76   Ryu and Won, 'The modern production'.

77   Korea Forest Service, *Leveraging Public Programmes with Socio-economic and Development Objectives to Support Conservation and Restoration of Ecosystems: Lessons Learned from the Republic of Korea's National Reforestation Programme* (Daejeon, Korea Forest Service, 2014), www.cbd.int/ecorestoration/doc/Korean-Study_Final-Version-20150106.pdf.

78   Scott Yorko, 'What could Americans learn from the way Koreans hike?', *Backpacker*, 26 March 2024, www.backpacker.com/stories/adventures/group-hiking-in-south-korea.

79   David Slatter, 'North Korea's mountain man goes hiking in the axis of evil', *NK News*, 6 February 2013, www.nknews.org/2013/02/north-koreas-mountain-man-goes-hiking-in-the-axis-of-evil.

80   Mason, 'The background'.

81   Yeong-Seok Jo and John Thomas Baccus, 'Are large cats compatible with modern society on the Korean Peninsula?', *Ecological Restoration*, vol. 34, no. 3 (2016), pp. 173–83.

82   Lee, 'Trekking on sacred'.

83   David A. Mason, *The Korean Forest Culture of the Baekdu Daegan: Spiritual and Folk Heritages along Korea's Grand 'Tiger's Spine' Mountain-System* (Daejeon, Korean Forest Service, 2011).

84   Mi Yoon Chung, Jordi López-Pujol and Myong Gi Chung, 'Is the Baekdudaegan "the Southern Appalachians of the East"? A comparison between these mountain systems, focusing on their role as glacial refugia', *Korean Journal of Plant Taxonomy*, vol. 46, no. 4 (2016), pp. 337–47; Mason, *The Korean Forest Culture*.

85   'North Korea seizes South's Mount Kumgang resort assets', BBC News, 22 August 2011, www.bbc.com/news/world-asia-pacific-14611873.

86   Park and Kyle, 'Identifying place meanings'.

87   Ryu and Won, 'The modern production'.

88   Peter Hayes, 'Building on Baekdudaegan: Peacemaking through ecological restoration', *Global Asia*, vol. 14, no. 4 (2019), pp. 91–7.

89   Chi-Young, 'The people'.

Epilogue

1   'World's Fair projects', *Chicago Tribune*, 2 November 1889, p. 9.

2   William Gilpin, *Cosmopolitan Railway: Compacting and Fusing Together all the World's Continents* (San Francisco, CA, History Company, 1890).

3   'Project summary', InterContinental Railway (ICR), 2018, https://static1. squarespace.com/static/58b72801e3df2821397e13cb/t/5b62c10003ce64ad 355d9967/1533198598103/ICR+Project+Summary+July+2018.pdf.

4   Euan Hague, 'More imagined than real: The Jefferson Davis Highway', *Journal of the Society for Commercial Archaeology*, vol. 28, no. 2 (2010), pp. 14–19.

5   Linda Poon, 'The racial injustice of American highways', Bloomberg, 3 June 2020, www.bloomberg.com/news/articles/2020-06-03/ what-highways-mean-to-the-george-floyd-protesters.

Acknowledgements

I am indebted to the many people who have helped fashion this book from concept into reality. In the first instance, I am grateful to Andrew Nurnberg and the dynamic team at Andrew Nurnberg Associates for trusting both in the nascent idea of *Earth Shapers* and in my previous book, *Invisible Lines*. Special thanks go to my outstanding agent Michael Dean for guiding me through my first years in the trade publishing world, offering unwavering support and insightful advice at every step along the way. A huge thank you to the indefatigable and delightfully inquisitive Profile Books team: to Andrew Franklin for seeing potential in my work as a new author; to Georgina Difford for managing the editorial process with great discernment; to Zara Sehr Ashraf for energetically supporting the book; and to Robert Loyko-Greer for shrewdly overseeing the publicity campaign.

Enormous thanks are also due to Robert Davies for meticulously copy-editing the manuscript and patiently advising me on what must have seemed like the most trivial of points, Hilary Bird for scrupulously compiling the book's index, and Dominic Beddow for conveying the complexities of our earth-shaped geography in such elegant and enlightening maps. I am deeply appreciative of Sinem Erkas for designing the handsome and commanding jacket, and the marketing/rights teams at both Andrew Nurnberg and Profile for sculpting interest in the book in the United Kingdom and beyond. Most of all, I am grateful to Izzy Everington, my phenomenal editor, for providing judicious suggestions and unswerving guidance from the day I started writing *Earth Shapers*. From forging the title to streamlining the notes, Izzy's role and impact are truly inscribed on every page.

Conducting research for *Earth Shapers* has been fun, enriching and often challenging. I would like to thank Euan Hague for helping to cultivate my fascination with Chicago's geography and for actively pointing me in the direction of different pieces of evidence of the city's Indigenous history, as well as for introducing me to several more modest examples of earth shaping touched on in this book. Thank you to Miranda Melcher for insightfully suggesting that I explore Mozambique's distinctive railway infrastructure, immediately following an interview with New Books Network. I am also appreciative of my students at DePaul University for piquing my interest in various of the world's distinct and distinctive places, some of which appear within the pages of *Earth Shapers*.

Last but not least, I owe boundless gratitude to my immediate family: my father, Alan, my mother, Leone, my brother, Nathaniel, and my wife, Eleanor. In your own ways, you have all sparked and sustained my captivation with the world around us, and motivated me to search ever more deeply for the narratives and concerns written on our planetary palimpsest.

Index